Place Event Marketing in the Asia Pacific Region

This book explores the fascinating phenomenon of place event marketing in the Asia Pacific region. It examines procedures in the promotion and branding of places that use events to shape their identities.

It considers how events are used in forming a branded image of a place and disseminate information about it. This innovative book offers theoretical insights of the opportunities and challenges related to place event marketing. With contributions from leading thinkers in the field, chapters also draw on empirical examples to showcase a variety of events across the Asia Pacific, such as MICE, sporting events, festivals, and religious and cultural celebrations. The book explores the importance of such events for the socio-economic development of urban regions. Today, the Asia Pacific is one of the world's fastest developing regions and its rising economic power is accompanied by the growing importance of the tourism and event sector. The book is a unique study relating to a very exceptional region of the world. The role of events in tourism development and the rise of the region's soft power is presented through carefully selected examples of cities from different countries. The book concludes with commentary on the future directions for research in this area.

Written in an accessible style, this book will be of great interest to students, scholars, and practitioners working in events studies, urban studies, tourism, place branding and promotion, business and management studies, geography, sociology, and sport and leisure studies.

Waldemar Cudny is an Associate Professor in Faculty of Geographical Sciences at The University of Lodz, Poland. He specialises in urban, tourism geography, and event studies. His publications include such books as *City Branding and Promotion: The Strategic Approach* and *Urban Events, Place Branding and Promotion: Place Event Marketing*.

Routledge Contemporary Perspectives on Urban Growth, Innovation and Change

Series edited by Sharmistha Bagchi-Sen, Professor, Department of Geography and Department of Global Gender and Sexuality Studies, State University of New York-Buffalo, Buffalo, NY, USA and **Waldemar Cudny**, Associate Professor. Working at the Faculty of Geographical Sciences at The University of Lodz, Poland.

Urban transformation affects various aspects of the physical, social, and economic spaces. This series contains monographs and edited collections that provide theoretically informed and interdisciplinary insights on the factors, patterns, processes and outcomes that facilitate or hinder urban development and transformation. Books within the series offer international and comparative perspectives from cities around the world, exploring how 'new life' may be brought to cities, and what the cities of future may look like.

Topics within the series may include: urban immigration and management, gender, sustainability and eco-cities, smart cities, technological developments and the impact on industry and on urban societies, cultural production and consumption in cities (including tourism, events and festivals), the marketing and branding of cities, and the role of various actors and policy makers in the planning and management of changing urban spaces.

If you are interested in submitting a proposal to the series please contact Faye Leerink, Commissioning Editor, faye.leerink@tandf.co.uk.

City Branding and Promotion
The Strategic Approach
Waldemar Cudny

Urban Events, Place Branding and Promotion
Place Event Marketing
Edited by Waldemar Cudny

Place Event Marketing in the Asia Pacific Region
Branding and Promotion in Cities
Edited by Waldemar Cudny

Place Event Marketing in the Asia Pacific Region

Branding and Promotion in Cities

Edited by Waldemar Cudny

Routledge
Taylor & Francis Group

LONDON AND NEW YORK

First published 2021
by Routledge
2 Park Square, Milton Park, Abingdon, Oxon OX14 4RN

and by Routledge
605 Third Avenue, New York, NY 10158

Routledge is an imprint of the Taylor & Francis Group, an informa business

British Library Cataloguing-in-Publication Data
A catalogue record for this book is available from the British Library

Library of Congress Cataloging-in-Publication Data
A catalog record has been requested for this book

ISBN: 978-0-367-42354-4 (hbk)
ISBN: 978-1-032-06101-6 (pbk)
ISBN: 978-0-367-82376-4 (ebk)

Typeset in Times New Roman
by Deanta Global Publishing Services, Chennai, India

Contents

Figures

Tables

Contributors

Christopher Barron is an experienced consultant with a demonstrated history of working in the sport and active recreation sector, sales and marketing, as well as services industry. Chris has worked with a number of national sports organisations, including New Zealand Rugby and Sport New Zealand. In 2020, Chris graduated with a Master of Business specialising in Sport Leadership and Management from Auckland University of Technology (AUT). He is currently a teaching and research assistant within AUT's School of Sport and Recreation.

Waldemar Cudny specialises in urban and tourism geography. Currently, he holds the position of Associate Professor at The University of Lodz, Faculty of Geographical Sciences (Poland). He investigated festivals and their impacts on cities, the socio-economic transformation of post-socialist cities, urban tourism, and car tourism. He has published a number of articles and several books. In 2019 his latest book titled *City Branding and Promotion: The Strategic Approach* (Routledge, 2019) was published. In 2020 he edited a book titled *Urban Events, Place branding and Promotion. Place Event Marketing* (Routledge 2020), where several case studies from five continents were analysed, presenting event marketing strategies in cities.

Aparajita De, is an Assistant Professor at the Department of Geography, Delhi School of Economics, University of Delhi. Her research largely falls within the scope of urban and media studies. Aparajita has been working persistently on the theme of imaginations and constructions of space in everyday lives. She has conducted research and published on the socio-spatial construction of Hindu–Muslim identities in the backdrop of communal violence; notions and practices of urbanisms in South Asia vis-à-vis understandings of modernity, particularly within the current post-colonial and decolonial discourses. Aparajita's current project involves the mapping of the colonial heritage and making of the colonial past of Kolkata in new media.

Daichi Oshimi is an Assistant Professor in the School of Physical Education at Tokai University, Japan. His area of expertise is in the socio-economic aspects of the sport. His grant-supported research focuses essentially on the impacts and leveraging of various sized sports events and their meaning for host communities and tourism. He also studies sports consumer behavior in a wide variety of fields. Dr. Oshimi has published his work in high-quality journals of sport management and event studies. He is deputy chief editor of the Asian Sport Management Review.

Sanjukta Sattar is an Associate Professor of Geography at the University of Mumbai, India. Her academic interest has been to examine the effectiveness and impact of tourism on regional development and studying various issues and challenges related to urban development.

Carolin Schlueter (MIB, Rikkyo University) is a doctoral student of international sport business in the Graduate School of Business at Rikkyo University in Tokyo, Japan. Her research interests are the effectiveness of sport sponsorships.

Shu-Wei Tsai is a Ph.D.candidate in the Geography Department at the University of California at Berkeley. Her research interests are urbanisation, heritage preservation, cultural landscape, urban modernisation, and urban culture. The regional focus is China and Taiwan. Her dissertation project is the heritagisation of a World Heritage site – the Chinese Grand Canal in contemporary social and cultural contexts of Chinese cities.

Yosuke Tsuji (Ph.D., Texas A&M University) is Professor of Marketing in the Department of Global Business at Rikkyo University in Tokyo, Japan. His research interests are branding in sports and sports event marketing effectiveness. His research has been published in the *Journal of Sport Management, Sport Management Review*, and *Sport Marketing Quarterly*, among others.

Nicholas Wise is an Assistant Professor in the School of Community Resources & Development at Arizona State University. Dr. Wise completed his Ph.D. in Human Geography at Kent State University in Ohio and, upon completion, he worked as a Lecturer in International Sport, Events, and Tourism Management at Glasgow Caledonian University from 2012 to 2015. His current research focuses on competitiveness, social regeneration, and community impacts, conducting work across the areas of sport, events, and tourism. He has edited and released nine books, most recently *Events, Places and Societies* (Routledge, 2019, with J. Harris) and *Tourism, Cultural Heritage and Urban Regeneration* (Springer, 2020, with T. Jimura).

Richard Keith Wright is a Senior Lecturer within the Department of Sports Leadership and Management. His teaching, research and industry engagement activities focus on active lifestyle entrepreneurship and event management as a tool for social development and positive well-being. Dr. Wright currently sits on the executive committee of Auckland Sunday Football Association, the AUT Centre for Active Ageing and the Australian and New Zealand Association for Leisure Studies. He is an associate editor of the *International Journal of the Sociology of Leisure* and the *Annals of Leisure Research*.

Yifan Xu is a Ph.D. candidate in the Department of Arts Administration, Education and Policy at the Ohio State University. With the background in the cultural industry management, arts marketing, and cultural policy studies, she is interested in the interdisciplinary exploration of arts and cultural events. Particularly, she focuses on the dynamics involved in the making of international performing arts festivals that serve various stakeholders and agendas in a changing social, economic and political environment. Previously studying in Shanghai for seven years, she had experiences with the booming event sector and creative industries in the city. She studies the China Shanghai International Arts Festival as a case for her dissertation on festival resilience from the perspective of legitimacy.

1 Introduction

Waldemar Cudny

The scope of the book

This edited volume is devoted to the role of different types of events in the promotion and branding of urban spaces. This is the second volume dedicated to this issue published by Routledge. The first volume was edited by Professor Waldemar Cudny and published by Routledge in 2020 under the title *Urban Events, Place Branding, and Promotion: Place Event Marketing.*

Whereas the previous publication encompassed case studies from five continents (North and South America, Europe, Asia, and Africa) this book focuses on cities from the Asia-Pacific region. The main research aims of this volume include:

- discovering how different kinds of events could be successfully used in place event marketing strategies encompassing promoting and branding of different types of urban places with the use of events,
- analysing how various public and non-governmental institutions can implement place event marketing activities in cities,
- investigating the effects of place event marketing activities on urban place promotion and branding,
- presenting a variety of empirical case studies with lessons regarding the use of different events in the promotion and branding of urban places from the Asia-Pacific region.

Methods

The book contains eight chapters with case studies presenting events organised in different countries of the studied region. Various types of research methods were used in these studies. The research method is a way of thinking and bringing to life solutions and new ideas that have scientific value (Sankey 2008). Whereas according to Apanowicz (2002, p. 59) the method of scientific research is "a set of theoretically justified procedures conceptual and instrumental, covering the entirety of the proceedings research aimed at solving a specific scientific problem". Methods may be divide into several categories; however, one of the most common divisions is based on the research procedure and type of materials being used. Here we can distinguish quantitative and qualitative methods. The first group uses numerical data obtained from empirical studies and measurements. The latter covers methods based on the analysis of the perception-based studies based on the opinions of other people, interviews, and the researcher's observation (Cudny 2016).

Both types of methods were used in this book. Some of the chapters were based on qualitative methods like interviews and observation, as well as the ethnography method and the analysis of posts and comments published on social media and blogging platforms (see chapters on Hangzhou, Shanghai, ASEAN countries, Kolkata, Mumbai, and Auckland). Some chapters used questionnaires and their quantitative analysis with the use of statistical methods (see chapters on Okinawa and Saitama).

Theoretical background

This book is devoted to events; therefore, defining the issue is a key element for further analysis. Definitions of events may be traced through different publications (Hall 1989; Goldblatt 1990; Getz 1991; Tara-Lunga 2012). According to Getz (2008, 404)

> Planned events are spatial-temporal phenomena, and each is unique because of interactions among the setting, people, and management systems – including design elements and the program. Much of the appeal of events is that they are never the same, and you have to be there to enjoy the unique experience fully; if you miss it, it's a lost opportunity.

Events are very diversified and thus may be divided into different types according to several criteria (see Jago and Shaw 1998; Arcodia and Robb 2000; Cudny 2016).

A comprehensive division of events was presented in the publication of Getz (2008). The author divided planned events according to their size and scale into

- local events (periodic and one-time);
- regional events (periodic and one-time);
- periodic hallmark events (high tourist demand and high value);
- occasional mega-events (high tourist demand and high value).

Getz (2008, 404) also divided planned events according to the theme and distinguished the following groups:

1. Cultural celebrations encompassing festivals, carnivals, commemorations, and religious events.
2. Political and state events including summits, royal occasions, and VIP visits.
3. Arts and entertainment encompassing concerts and award ceremonies.

4. Business and trade including meetings, conventions, consumer and trade shows, fairs, and markets.
5. Educational and scientific events, including conferences, seminars, and clinics.
6. Sports competitions including amateur/professional and spectator/participant events.
7. Recreational events, encompassing sport or games for fun.
8. Private events including weddings, parties, and social gatherings.

Urban spaces across the globe compete to gain resources such as investors, investments, immigrants, and tourists who bring money into the economy of a visited destination. Therefore different techniques are used in destination management, including marketing management. As defined by Kotler and Armstrong (2010, 29) marketing is "a social and managerial process by which individuals and organisations obtain what they need and want through creating and exchanging value with others. In a narrower business context, marketing involves building profitable, value-laden exchange relationships with customers".

Marketing management arises from marketing and may be treated as the procedure of leading an institution (i.e. taking advantage of the organisation's resources, planning actions, implementing and controlling them) in order to achieve its goals by creating a beneficial exchange between an institution and its customers. When marketing is applied to places it is called place marketing, and city marketing when it refers to cities (Cudny 2020a, p. 13–14).

Kotler and Armstrong (2010, 252–253) defined place marketing as "Activities undertaken to create, maintain or change attitudes or behaviour towards particular places. Cities, states, and even nations compete to attract tourists, new residents, conventions, and company offices and factories".

Cities (as one of the places undergoing place marketing activities) want to attract the above-mentioned resources and customers who would like to consume their place products (i.e. urban investments land, tourist assets, services, and residential areas). Therefore cities introduce place marketing activities. Thanks to marketing efforts the offer of cities is strengthened and new services, attractions, and other urban "products" appear. They are tailored according to consumers' needs and properly managed and promoted in order to attract consumers to a given city. Place marketing encompasses product creation, its management, promotion, and distribution (Cudny 2020a).

Event marketing is included in marketing procedures and in a business context, it may be defined as "creating a brand marketing event or serving as a sole or participating sponsor of events created by others. The events might include anything from mobile brand tours to festivals, reunions, marathons, concerts, or other sponsored gatherings. Event marketing is huge, and it may be the fastest-growing area of promotion" (Kotler and Armstrong, 505). Firms are involved in event marketing because it is the way they can promote their products, reach potential customers in order to learn about their needs, exchange information, and form relationships with them, as well as creating a brand image of a company and their products.

However, in the last years, event marketing is well applied to places such as cities and towns. Event marketing is, in this case, part of city marketing and city branding activities. As stated by Cudny (2020b, 14)

> City marketing involves urban socio-economic development and promotion. The main aim of promotion is to create communication between an institution and the end-users of its offer (consumers). Among the basic means of promotion which form the so-called promotional means, there are advertising, sales promotion, public relations personal selling and direct marketing.

Place branding is another term connected with place event marketing. It is a complex and multidimensional policy, which is introduced in cities for different purposes. It aims at the socio-economic development of cities due to long-term investments and attraction of newcomers and tourists, creation of a well perceived urban brand. However, place branding in cities is also a result of the influence of different interest groups and stakeholders operating in the urban space (like firms, politicians, social groups, etc.). Branding policy results in actions and patterns that change the social and economic structures, as well as the physical space and image of a city, and creates its brand (Lucarelli 2018).

The concept of place event marketing is central to this edited volume. The issue had been presented by Cudny (2020b). The author defined the scope of event marketing, its connections to city marketing, branding, and urban promotion. According to Cudny (2020b), place event marketing in cities is a twofold procedure. On the one hand place event marketing is a city development policy, while on the other hand it is a means of city promotion (Figure 1.1.).

Place event marketing regarded as a city development policy is similar to place or city branding. Events are enriching the urban product, creating tourism, and are part of the tourism industry. Therefore they generate income for a host city and create jobs. Moreover, many of them (e.g. sports mega-events like the Olympics) generate large-scale investments (stadiums, transportation facilities, etc.). These investments also generate economic development and develop the product of a certain urban place. Moreover, events found social opportunities; for instance, local communities may get involved in event organisations, as volunteers, advisors, participants (e.g. local artists engaged in festival creation), and local firms can participate in event creation by delivering services, and thus people can get jobs and revenue. Event facilities can host events in the subsequent years, and the infrastructure may be used by inhabitants and tourists long after the event is gone.

Broadly speaking, properly prepared and managed event creation can generate a socio-economic and infrastructural boost which may be seen as a long-term development plan. The perfect realisation of an event-led regeneration and development strategy of this kind was the Summer Olympics in Barcelona in 1992. The sports mega-event was used in the capital of Spanish Catalonia as an urban regeneration tool. The preparations for the Olympics started in the 1980s with the elaboration of a complex infrastructural and economic development plan. Thanks to its introduction the post-industrial, harbor city has been transformed, and after

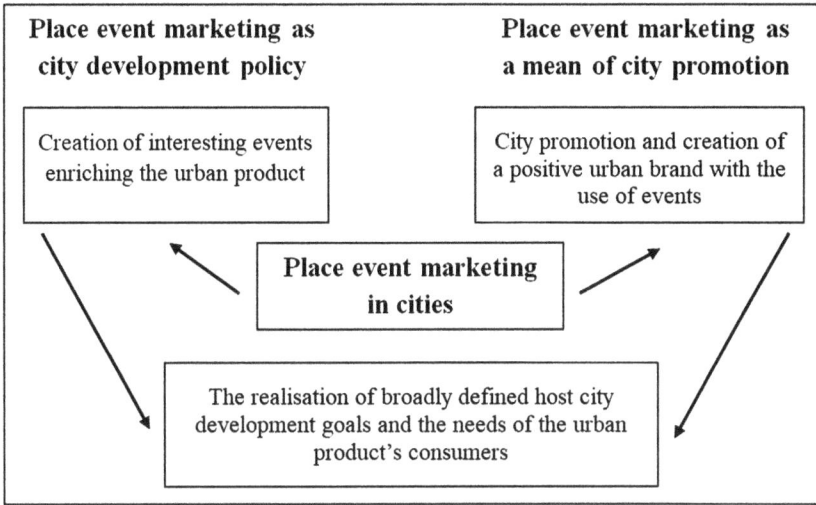

Figure 1.1 The structure of place event marketing in cities. Source: Own elaboration on the basis of Cudny (2020b, 17).

the 1992 Olympics, Barcelona became a vibrant urban centre with a strong focus on services, culture, events, and tourism. This development plan based on event regeneration was so successful that it is often referred to as the Barcelona model (see: Garcia-Ramon and Albet 2000).

Place event marketing understood as a means of city promotion is no less important (Cudny 2020b). Events are a platform for city promotion. Mega and hallmark events are widely presented in the traditional media (TV, newspapers), in social media (FB, Twitter, Instagram), and by word of mouth. Regional or even small-scale events receive their media coverage too. Moreover, many events (e.g. concerts, festivals, and sports events) are attended by celebrities (sportsmen, actors, musicians) who are widely known. Therefore the media is eager to broadcast these events and mention host cities at the same time.

Both aforementioned issues (i.e. event marketing as an urban development plan and as a promotional tool) allow the realisation of broadly defined development goals of cities (i.e. creation of socio-economic growth, tourism development, the growth of infrastructure, brand creation, and image strengthening). For example, if a city struggles to attract tourists, development of events may be a factor attracting them and creating demand for services that may drive the growth of tourism investments and the tourism industry. If a city requires more investments in infrastructure and hallmark objects (concert halls, stadiums) the organisation of hallmark or mega-events is a perfect opportunity to gain such facilities. All these benefits are strengthened when events promote the destination while being broadcasted in the media or promoted via word of mouth. This helps to develop the socio-economic conditions of a city and strengthen its image and brand (Cudny 2020b) (Figure 1.1.).

The Asia Pacific region

The book focuses on the Asia Pacific region, currently one of the fastest developing areas of the world. It encompasses a large group of countries located in or close to the Western Pacific Ocean and includes countries from East and South Asia, Southeast Asia, and Oceania (Figure 1.2). The region is referred to as APAC, Asia-Pac, or AsPac. It started to gain international attention in the 1980s when the Asian economy boosted. At that time, the area at the fringe of Asia, Oceania, and the Western Pacific Ocean increased its economic and political importance (http s://worldpopulationreview.com/country-rankings/apac-countries).

International organisations define the boundaries of the Asia Pacific region differently. For instance, the International Energy Agency (IEA) lists 23 countries in the Asia Pacific region inhabited by 4.1 billion people (https://www.iea.org/regions/asia-pacific). According to the Food and Agriculture Organization of the United Nations (FAO), the region under study encompasses 49 countries from Asia and Oceania included in the organisation's forestry study (http://www.fao.org/3/x2613e/x2613e04.htm) (Table 1.1).

According to FAO, the Asia Pacific Region covers 2.8 billion hectares of land (ca. 22 percent of the global land area). It ranges from the borders of China–Mongolia in the north to the southern tip of Australia and New Zealand in the south. This region covers the broad spectrum of natural conditions, including northern temperate and boreal through the tropical and subtropical zones, from the high plateaus and mountains of Pakistan to the islands of North Asia and the South Pacific (http://www.fao.org/3/W4388E/w4388e03.htm) (Table 1.1).

United Nations World Tourism Organization (UNWTO) included 48 countries in the Asia Pacific region. These countries are divided into the following subregions of North-East Asia, South-East Asia, Oceania, and South Asia (Table 1.1).

The case studies incorporated in this book encompass events from cities located in four large Asia Pacific countries, like China (Hangzhou and Shanghai), India

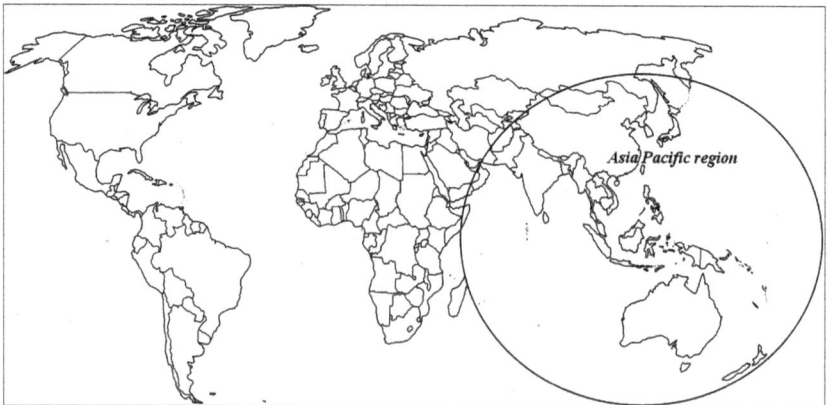

Figure 1.2 The Asia Pacific region on the map of the World. Source: Own elaboration.

Table 1.1 Countries included in the Asia Pacific region according to IEA, FAO, and UNWTO

Division of the Asia Pacific region	The list of countries
According to the International Energy Agency (IEA)	People's Republic of China, India, Japan, Korea, Indonesia, Thailand, Australia, Chinese Taipei, Pakistan, Malaysia, Vietnam, Philippines, Bangladesh, Singapore, Myanmar, New Zealand, Democratic People's Republic of Korea, Hong Kong, Nepal, Sri Lanka, Cambodia, Mongolia, Laos, Brunei Darussalam
According to the Food and Agriculture Organization of the United Nations (FAO)	**Asia:** Bangladesh, Bhutan, Brunei Darussalam, Cambodia, China Mainland, Taiwan, Hong Kong SAR, India, Indonesia, Japan, Korea, Democratic People's Republic of Korea, Laos, Malaysia, Maldives, Mongolia, Myanmar, Nepal, Pakistan, Philippines, Singapore, Sri Lanka, Thailand, Vietnam, **Oceania:** American Samoa, Australia, Christmas Island, Coco (Keeling) Islands, Cook Islands, Fiji Islands, French Polynesia, Guam, Kiribati, Federal States of Micronesia, Northern Mariana Island, Nauru, New Caledonia, New Zealand, Niue, Norfolk Island, Pacific Islands Trust Territory, Palau, Papua New Guinea, Samoa, Solomon Islands, Tonga, Tuvalu, Vanuatu, Wallis and Futuna Islands
According to the United Nations World Tourism Organization (UNWTO)	**North-East Asia:** China, Hong Kong (China), Japan, Korea (DPRK), Korea (ROK), Macao (China), Mongolia, Taiwan, **South-East Asia:** Brunei, Cambodia, Indonesia, Laos, Malaysia, Myanmar, Philippines, Singapore, Thailand, Timor-Leste, Vietnam, **Oceania**: American Samoa, Australia, Cook Islands, Fiji, French Polynesia, Guam, Kiribati, Marshall Islands, Micronesia FSM, N.Mariana Islands, New Caledonia, New Zealand, Niue, Palau, Papua New Guinea, Samoa, Solomon Islands, Tonga, Tuvalu, Vanuatu, **South Asia**: Afghanistan, Bangladesh, Bhutan, India, Iran, Maldives, Nepal, Pakistan, Sri Lanka.

Source: Own elaboration on the basis of https://www.iea.org/regions/asia-pacific; http://www.fao.org/3/x2613e/x2613e04.htm; International Tourism Highlights 2019.

(Kolkata and Mumbai), Japan (Saitama, the prefecture and city of Okinawa), supplemented by a case study of New Zealand sports event (Auckland). Moreover, one case study presents an overview of several cities from countries included in the Association of Southeast Asian Nations (ASEAN). The ASEAN organisation was established in 1967 and aims for regional socio-economic development. ASEAN encompasses the following countries: the Philippines, Indonesia, Malaysia, Singapore, Thailand, Brunei, Vietnam, Laos, Myanmar, and Cambodia.

All countries included in this edited volume are part of the Asia Pacific region according to the IEA, FAO as well as UNWTO divisions (Table 1.1.).

This book does not intend to present a full and comprehensive evaluation of place event marketing in the cities of the Asia Pacific region. The book aims to present selected examples and start a discussion on the role of events in urban development, branding, and promotion in this region. The examples of countries and cities presented in this publication form an introduction to further comprehensive research on this issue.

The rapid economic growth of the Asia Pacific began at the turn of the 1980s and 1990s. At that time, significant socio-economic and political changes took place in countries in this area. In many of them, an intense system change occurred, encompassing the democratisation of social and political lives. Moreover, these countries have started to implement complex economic development strategies. They were supposed to bring the level of development and living standards closer to that of highly developed countries. Some of the Asia Pacific countries tried to grow an economy based on the export of their natural resources like oil (Indonesia, Malaysia). While other countries (mainly those poor in raw materials) tried to found their growth strategies on the development of technologically advanced branches of production. Additional, but not less important, factors stimulating the region's growth included the liberalisation of international trade, the development of globalisation and, the influx of foreign investors possessing know-how and investment capital (Morley and Ichimura 2015).

The Asia-Pacific countries had vastly contributed to the global economy in the last decades. Therefore the area is now one of the most important regions for contemporary global growth. For example, the share of GDP in emerging and developing Asian economies increased from 9% to 29% between 1980 and 2014. In 2012 the Asia-Pacific region itself contributed to 27.4% of the global GDP and to ca. 40% of global growth (Fang and Chang 2016).

The rising economic importance of the region was accompanied by its growing contribution to global tourism. According to the United Nations World Tourism Organization (UNWTO) statistics, 348 million tourists visited the Asia and the Pacific regions in 2018, and tourism revenues in this area reached $ 435 billion. The region of Asia and the Pacific in 2018 was second in the world after the European region in terms of the volume of tourist arrivals and income from tourism (International Tourism Highlights 2019).

In the last decades, the Asia Pacific region was the fastest-growing tourism region in the world. The tourism industry plays a significant role in many Asia Pacific countries. The results of tourism development in the region include economic growth, development of workplaces, and income as well as growth in leisure opportunities (Singh 1997; Tolkach et al. 2016).

According to the industry publications and literature review presented by Tolkach et al. (2016), one of the vital elements creating tourism in the region was the growing role of events, including festivals, sports, and business events. The cited authors presented the review of the major tourism trends in 2014, resulting from the selected media analysis, regarding the Asia Pacific region. Among them (listed in this publication under number 17), there was a major tourism trend defined as *events*. According to Tolkach et al. (2016, 1084), "A growing number of sports events take place in the Asia Pacific region. Cultural festivals can be a draw for the growing number of cultural tourists in the region. Government involvement is often required for successful mega-events to take place due to long-term planning and infrastructure development ".

There is a growing body of literature devoted to the role of different events in tourism development in the region under study. The publications present sports events, including recent mega-events organised in the Asia Pacific (see: Kim et al. 2006; Zeng and Luo 2008; Singh and Zhou 2016); religious events and festivals (see: McKercher et al. 2006; Ryan and Gu 2010); MICE events (see: Mistilis and Dwyer 1999; Nadkarni and Wai 2007; McCartney 2008) as well as complex studies on the impact of events on tourism and regional development (see: Prideaux et al. 2013; Hassan and Sharma 2018). Some of the recent publications considered the role of events in the creation of image and branding of nations and countries in the Asia Pacific region (see: Berkowitz et al. 2007; Lee 2010; Bodet and Lacassagne 2012; Li and Kaplanidou 2013; Lai 2018).

This book tries to connect both of the research areas. On the one hand, it shows how various types of events organised in selected countries in the Asia Pacific region affect the development of tourism and create other socio-economic benefits to the cities and towns located there. On the other hand, the analysis presented in the book describes how the organisation of events contributes to changes in the destination's image and brand creation.

The COVID-19 impact

This book was written before the start of the COVID-19 pandemic. Coronavirus evolved when the text of the book was ready and was undergoing the editing process. Therefore the book does not analyze coronavirus implications for the tourism and event industry in the Asia - Pacific region. However, it is impossible not to refer to the COVID-19 pandemic and its impact on the shape of the modern tourism sector and the event industry.

As indicated by the United Nations World Tourism Organization (UNWTO) report published in the summer of 2020, as a result of the global coronavirus pandemic, the number of international tourist trips and income from tourism will significantly decline. Depending on the forecast variant presented by the UNWTO, the decline in tourist traffic in 2020 compared to the previous year is estimated at 58–78%. The crisis is particularly vulnerable to small businesseses, which are responsible for 80% of global tourism (http://www.unic.un.org.pl/oionz/raport: -turystyka-i-covid-19/3373).

At the end of 2020, it was already clear that the UNWTO's estimates were accurate. The second wave of the coronavirus in the fall of 2020 brought a high death toll in most countries of the world. It was associated with the closure of many economies (lockdowns) and the closure of air travel. It also affected countries from the region described in the book, i.e. the Asia-Pacific region. As a result of the pandemic, the organisation of many mass events was forbidden and, among others, the 2020 Summer Olympics in Tokyo was moved for the summer of 2021.

Certainly, the short-term effects of the pandemic for the tourism and event sectors will be very unfavorable. A decline in tourist traffic, income from tourism, the bankruptcy of many companies (especially small- and medium-sized ones), including companies involved in the production of events, is inevitable. In addition, many events, even mega-events planned for 2020, have been canceled. Therefore, the destinations where they were to take place will not achieve the planned income, and expected promotional goals, however, will pay the costs of many years of preparation. In some cases, the positive impacts may be achieved in the subsequent years (if the event can be moved). However, sometimes (e.g. in the case of annual events) the benefits of their 2020 editions will be lost.

On the other hand, the end of 2020 brought hope for normalisation, mainly due to the development of several very effective vaccines against the coronavirus. In addition, economic data indicated the possibility of a quick recovery of the economy after the shock related to the coronavirus.

Certainly, we will see a new social and economic reality in the coming years. However, as long as the vaccine is effective, we will undoubtedly deal with the recovery of the economy after the coronavirus crisis. Consequently, cities and countries will be looking for opportunities to reactivate the tourism sector after the pandemic. One of the paths will be the development of events. They can be an important element in restoring the tourism economy and promoting cities, regions, and entire countries. Undoubtedly, people will continue to search for the unusual experiences that events provide. In addition, after the lockdown period, they will also look for contacts with other people and meeting places with attractive cultural and entertainment programs. Therefore, it should be expected that tourism and the event sector will experience a revival. This will particularly apply to the Asia-Pacific region discussed in the book, which is far better at dealing with the coronavirus pandemic than the rest of the world. On the other hand, despite the introduction of the vaccine, it will still be necessary to follow sanitary procedures and this will be a big challenge for mass event organisers in the post-pandemic period.

The structure of the book

The book starts with the introductory chapter, which is followed by chapters with case studies representing events organised in a selected group of countries from the Asia Pacific region. Chapter 2, written by Shu-Wei Tsai, discusses the branding strategies introduced in Hangzhou (China) on the case study of the G20 Hangzhou Summit. It presents the mega-event impacts on tourism, urban renewal, and infrastructure development, and on the social sphere of the city of Hangzhou. Moreover, the author also presents the impacts of the event on city brand and promotion and its perception by citizens.

Chapter 3 authored by Yifan Xu presents the role of the China Shanghai International Arts Festival in the branding of the city of Shanghai. The studied event has a diverse program encompassing performances and exhibitions, arts education events, trade fairs, forums, and public events. Through them, CSIAF communicates different messages regarding Shanghai to its local audience and international participants. Drawn on festival stakeholders' marketing strategies and the images of Shanghai framed in CSIAF-related promotions, this chapter explores how the city of Shanghai is branded by its various stakeholders as a Global City.

Chapter 4 by Nicholas Wise represents a different approach. Instead of focusing on one specific case study, it presents a comparison of the event-led branding and promotional strategies in several cities from the Asia Pacific region, united in the ASEAN organisation (the Association of Southeast Asian Nations). Nicholas Wise presents how ASEAN countries use large-scale events to build international influence, promote cultural values, and encourage business development. Several examples are presented by the author to characterise the development of events in ASEAN alongside conceptual discussions of regeneration, place image, and competitiveness to address product-oriented and value-added perspectives of destination marketing and branding in Asia Pacific.

Chapters 5 and 6 describe case studies of sports events held in different cities of Japan and their impacts on the cities' image and brand. Moreover, unlike the rest of the chapters, they utilise quantitative methods i.e. statistical analysis of the results of questionnaires distributed among event participants. The first of these chapters, written by Yosuke Tsuji and Carolin Schlueter, presents the brand association of Okinawa Marathon and its effects on places. The second chapter, authored by Daichi Oshimi is devoted to the socio-economic and branding impacts of the Le Tour de France Saitama Criterium on the Japanese city of Saitama. The author presented how the sports event is organised, and what kind of economic, social, and branding impacts does it have on a host city and the Saitama region.

The subsequent chapters (Chapters 7 and 8) represent case studies from India. Chapter 7 written by Aparajita De is devoted to Durga Puja, a religious festival held in Kolkata. The chapter investigates how local heritage and culture intersects with the place branding of the city, in particular how social media circulates this brand image of the city. The branding of the city builds on the perception of the festival's collective participation, which cuts across different social categories of caste and religion, and a spectacular exhibition of Bengali culture.

Chapter 8 by Sanjukta Sattar characterises the cultural and event sector of Mumbai. This chapter is based on media analysis and in-depth interviews. The city image as the film, fashion, and entertainment capital, as well as a creative city, is presented in the chapter. Later the chapter analyses what roles these activities play in creating and re-positioning the image and branding of the city of Mumbai.

Chapter 9, written by Richard Keith Wright and Christopher Barron, presents the scheduled sports event America's Cup regatta which will take place in Auckland in 2021. The chapter characterises a unique opportunity for promoting the geographically isolated city of Auckland and the country of New Zealand to the world with the use of a world-famous regatta. The authors try to predict the opportunities and benefits resulting from hosting the event and factors affecting the urban development and branding of Auckland and New Zealand.

The book ends with a conclusion where the main results are presented and analysed against the concept of place event marketing.

References

Apanowicz J. (2002). *Metodologia ogólna*. Wydawnictwo BERNARDINUM, Gdynia.

Arcodia, C., Robb, A. (2000). A future for event management: A taxonomy of event management terms. In: Allen, J., Harris, R., Jago, L.K., A.J. Veal (eds.) Events beyond 2000: Setting the Agenda Proceedings of Conference on Event Evaluation, Research and Education, Sydney, July 2000, pp. 154–160, Australian Centre of Event Management School of Leisure, Sport and Tourism University of Technology, Sydney.

Berkowitz, P., Gjermano, G., Gomez, L., Schafer, G. (2007). Brand China: Using the 2008 olympic games to enhance China's image. *Place Branding and Public Diplomacy*, 3(2), 164–178.

Bodet, G., Lacassagne, M.F. (2012). International place branding through sporting events: A British perspective of the 2008 Beijing Olympics. *European Sport Management Quarterly*, 12(4), 357–374.

Cudny, W. (2016). *Festivalisation of Urban Spaces: Factors, Processes and Effects*. Springer, Cham.

Cudny, W. (2020a). *City Branding and Promotion: The Strategic Approach*. Routledge, London/New York.

Cudny, W. (2020b). The concept of place event marketing: Setting the agenda. In: Cudny, W. (ed.) *Urban Events, Place Branding and Promotion. Place Event Marketing*. Routledge, London/New York.

Fang, Z., Chang, Y. (2016). Energy, human capital and economic growth in Asia Pacific countries: Evidence from a panel cointegration and causality analysis. *Energy Economics*, 56, 177–184.

Garcia-Ramon, M.D., Albet, A. (2000). Pre-Olympic and post-Olympic Barcelona, a 'model' for urban regeneration today? *Environment and Planning A*, 32(8), 1331–1334.

Getz, D. (1991). *Festivals, Special Events, and Tourism*. Van Nostrand Reinhold, New York.

Getz, D. (2008). Event tourism: Definition, evolution, and research. *Tourism Management*, 29(3), 403–428.

Goldblatt, J.J. (1990). *Special Events: The Art and Science of Celebration*. Van Nostrand Reinhold, New York.

Hall, C.M. (1989). The definition and analysis of hallmark tourist events. *GeoJournal*, 19(3), 263–268.

Hassan, A., Sharma, A. (Eds.) (2018). *Tourism Events in Asia: Marketing and Development*. Routledge, London.

International Tourism Highlights (2019). *UNWTO World Tourism Organization*, Edition 2019, available at: https://www.e-unwto.org/doi/pdf/10.18111/9789284421152, accessed on 18.08.2020.

Jago, L.K., Shaw, R.N. (1998). Special events: A conceptual and definitional framework. *Festival Management and Event Tourism*, 5(1–2), 21–32.

Kim, H.J., Gursoy, D., Lee, S.B. (2006). The impact of the 2002 World Cup on South Korea: Comparisons of pre-and post-games. *Tourism Management*, 27(1), 86–96.

Kotler, P., Armstrong, G. (2010). *Principles of Marketing*. Pearson, New York.

Lai, K. (2018). Influence of event image on destination image: The case of the 2008 Beijing Olympic Games. *Journal of Destination Marketing & Management*, 7, 153–163.

Lee, A.L. (2010). Did the Olympics help the nation branding of China? Comparing public perception of China with the Olympics before and after the 2008 Beijing Olympics in Hong Kong. *Place Branding and Public Diplomacy*, 6(3), 207–227.

Li, X., Kaplanidou, K. (2013). The impact of the 2008 Beijing Olympic Games on China's destination brand: A US-based examination. *Journal of Hospitality & Tourism Research*, 37(2), 237–261.

Lucarelli, A. (2018). Place branding as urban policy: The (im) political place branding. *Cities*, 80, 12–21.

McCartney, G. (2008). The CAT and the MICE: Key development considerations for the convention and exhibition industry in Macao. *Journal of Convention & Event Tourism*, 9(4), pp. 293–308.

McKercher, B., Mei, W.S., Tse, T.S. (2006). Are short duration cultural festivals tourist attractions?. *Journal of Sustainable Tourism*,14(1), 55–66.

Mistilis, N., Dwyer, L. (1999). Tourism gateways and regional economies: The distributional impacts of MICE. *International Journal of Tourism Research*, 1(6), 441–457.

Morley, J.W., Ichimura, S. (2015). Introduction: The Varieties of Asia-Pacific Experience. In: Morley, JW (ed.), *Driven by Growth: Political Change in the Asia-Pacific Region*, pp. 3–34. Routledge, London.

Nadkarni, S., Wai, A.L.M. (2007). MACAO'S mice dreams: Opportunities and challenges. *International Journal of Event Management Research*, 3(2), 47–57.

Prideaux, B., Timothy, D., Chon, K. (Eds.) (2013). *Cultural and Heritage Tourism in Asia and the Pacific*. Routledge, London.

Ryan, C., Gu, H. (2010). Constructionism and culture in research: Understandings of the fourth Buddhist Festival, Wutaishan, China. *Tourism Management*, 31(2), 167–178.

Sankey, H. (2008). Scientific method. In: M. Curd and Psillos, S. (eds.) *The Routledge Companion to Philosophy of Science* (pp. 248–258). Routledge, Lodnon/New York.

Singh, A. (1997). Asia Pacific tourism industry: Current trends and future outlook. *Asia Pacific Journal of Tourism Research*, 2 (1), 89–99.

Singh, N., Zhou, H. (2016). Transformation of tourism in Beijing after the 2008 Summer Olympics: An analysis of the impacts in 2014. *International Journal of Tourism Research*, 18(4), 277–285.

Tara-Lunga, M.O. (2012). Major special events: An interpretative literature review. *Management & Marketing*, 7(4), 759.

Tolkach, D., Chon, K.K., & Xiao, H. (2016). Asia Pacific tourism trends: Is the future ours to see? *Asia Pacific Journal of Tourism Research*, 21(10), 1071–1084.

Zeng, K., Luo, X. (2008). China's inbound tourist revenue and Beijing Olympic Games 2008. *China & World Economy*, 16(4), 110–126.

2 When ordinary life becomes extraordinary

The branding of Hangzhou, China, during the 2016 G20 Summit

Shu-Wei Tsai

Introduction

This chapter discusses the strategies used in city branding with the use of mega-events. Moreover, it also examines how events could be used in internal and external marketing of a city and how such marketing endeavours impact citizens. Mega-events and their preparation most often take place in cities and trigger investment, infrastructure projects, art programming (García, 2004), or urban regeneration (Smith, 2012). No matter whether the event is organised by the state or is locally driven, mega-events tend to create positive images of host cities, regions, or states (Arnegger & Herz, 2016; Broudehoux, 2019; Cudny, 2020).

Though mega-events are perceived as an influential one-time event that is impossible to be replicated (Getz, 2007), the host society can be examined before, during, and after the event as those events might be used as future-oriented strategies for city marketing. The preparation for the event may well strengthen infrastructure, venues, and improve the physical environment to attract visitors and potential investors and bring positive effects on urban development (Maennig & Du Plessis, 2009). Moreover, some event legacies can be used for social programs in developing countries (Gunter, 2014).

With the aid of communication technology, diffused images affect the perceptions of the audience and visitors during the event. Once the event is over, the remaining facilities, infrastructures, and renovated environments are physical legacies to the citizens of the host cities. Moreover, the hosting experience often consolidates local pride and identity. In this sense, mega-events are used as marketing strategies for city authorities to advertise their competitiveness both externally (to visitors) and internally (to their citizens) (Arnegger & Herz, 2016; Broudehoux, 2019; Cudny, 2020).

The central questions this chapter seeks to answer are:

1. How does a host city brand itself by hosting a mega-event and balance external and internal branding strategies?
2. How do citizens perceive internal and external event-branding effects?
3. How does the event organisation influence citizens' life?

This case study focuses mainly on how the internal branding of the event affected the everyday practices of ordinary people in the host city in China. The 4–5 September 2016 G20 Summit organised in the Chinese city of Hangzhou was chosen as a case study to answer these questions.

This research is designed as a qualitative study. Methods such as participatory observation, interviews conducted between 2016 and 2018, and discourse analysis from the newspapers, government documents, blogs, and online forums were used to collect information from local people, student volunteers of the G20 summit, and preservationists. As this event was highly sensitive in terms of national security, net-ethnography was incorporated to collect the information of local people, without the presence of the researcher.

Literature review

The planned events belong to the following categories: Cultural celebrations, political and state, arts and entertainment, business and trade, educational and scientific, sports competitions, recreational, private events (Getz, 2007, 2008). According to this division, the studied case, the G20 Summit, belongs to the group of political and state events. However, in terms of the content of this summit, this event also belongs to the category of business event and the sector of MICE: Meetings, Incentives, Conferences, Exhibitions (Mistilis & Dwyer, 2000).

This chapter defines city branding as a set of policies and strategies with marketing communication to identify, position, and create an image of a specific city/place (Anholt, 2008; Kavaratzis & Ashworth, 2005). Since a place brand is hard to be treated like a product brand, the theories and practices between corporate branding and place branding are different (Ashworth & Kavaratzis, 2009). City branding can be seen as a way with strategic management of urban territories that encourage growth, development, and tourism (Cudny, 2020).

This chapter directly discusses place event marking understood

> as all activities aimed at enriching the product of a given city by offering its consumers a well-chosen, interesting and diverse portfolio of events. This activity also includes the promotion of a city through the events. Place event marketing, in turn, enables marketing communication with the recipients of urban products and shapes the positive image and brand of a city.
>
> (Cudny, 2020, 16–17)

The existing literature on city branding through mega-events emphasises how political ideology and economic concerns dominate top–down branding strategies in mega-events. Following this thread, two categories are shown as external branding and internal branding.

External branding

Studies on external branding can be categorised into two main accounts. First, mega-events are used as an external branding strategy that showcases national progress to other countries. This discussion also relates to developing countries aiming to increase their international visibility, not only the city per se but the country as a whole, by a presentation of their political and economic performance to the world. For example, Joo et al. (2017) point out that in the past three decades, South Korea has branded itself through four events, each of them with a specific function to signal the political and economic transformation of the South Korean state. For example, the 1988 Summer Olympic Games was to declare the democratic transition from authoritarian governance, and the 1993 Daejeon Expo was to announce the new economy led by technological industries rather than export-oriented industries. Mega-events have similar tasks for a nation to demonstrate its technological or scientific advance or, more broadly speaking, its global visibility, as shown in South Africa in the 2010 FIFA World Cup (Knott et al., 2015). Broudehoux (2007) argues that Beijing took advantage of the chance to host the Olympics to show off its transition from a traditional city to a brand-new spectacular city to claim its global city status. The latter three examples show that states in developing countries used mega-events to show their political ambitions.

Secondly, an essential function for external branding aims to deepen economic competition by using "mega-event strategies" (Burbank et al., 2001). These strategies are implemented to boost the urban economy and involve more financial calculations to either attract foreign investment or trump other competing cities, depending on the varying social and political contexts (Cope, 2015). This function is essential as some host cities in a developmental phase try to expand their tax base and levels of public attention from outside investment. Event host cities compete to claim their leading positions in not only national but global levels to seek investment opportunities. City governments develop improved infrastructure and financial policies to attract investors and visitors to suit the preference of the local elites. In Pyeong Chang's case, hosting the 2018 Winter Olympic Games was the only chance for this poor city to grow, so it strived not to co-host this event because it did not want to share the event-based benefits with other cities (Joo et al., 2017). This type of strategy, however, is more usually dictated by politicians rather than citizens due to the latter's lower levels of influence. Clearly, two accounts of external branding are often overlapped and not exclusive to each other.

Internal branding

Studies that discuss internal branding tend to explain how the host entities solve the challenges of economic and infrastructural development image-making and social reality towards internal stakeholders, mostly inhabitants. The host entity seeks to manage themselves as an "eventful city", which is defined as a city which "uses a programme of events to strategically and sustainably support long-term policy agendas that enhance the quality of life for all" (Richards, 2015, p. 39). However, event-based city branding focuses on positive image-making to attract recipients, such as potential investors and visitors and imagines delightful experiences within territorial marketing (Cudny, 2016; Ooi, 2010) which excludes negative images and the externalities of the assigned territory. Also, new displays in modern mega-events tend to create and enhance some universal values, such as world peace and sustainability, even though this promotion does not always work. For example, Western countries expected China to improve its bad human rights record through the opportunity of hosting the 2008 Olympic Games (Lenskyj, 2008). However, the design and planning of this event did not consider the well-being of citizens and migrant workers in Beijing as large-scale demolition projects (Shin & Li, 2013). Those studies show that it is not feasible to erase social differences and attain a satisfying life for all since society is not a homogenous entity.

Broudehoux (2007) points out that image construction during Beijing Olympics included much on public education on urbanity and civility, despite the gap between this image, the expectations of visitors, and the reality that local people faced (Zhang & Zhao, 2009). Image-making in mega-events also involves the removal and minimisation of unpleasant environmental elements and risks. Braathen et al. (2016) indicate that the hosting of the Olympic Games, Rio de Janeiro 2016, connects to the urban development framework involved in the demolition of undesirable neighbourhoods. This finding resonates with the previous statement regarding the positioning of the city in that the *favelas* (informal poor settlement in Brazil) were largely removed in Rio for the Olympic Games in 2016 (Broudehoux, 2019). This was done to showcase the "turnaround" urban reform agenda of Rio, which had been losing out to Brasilia and Sao Paulo in the last half of the 20th century (Richmond, 2016). Besides, tension can also occur when social unrest takes place at such an eye-catching public gathering. Mega-events provide platforms for protest, and in some cases, protestors challenge the very nature of these events (Timms, 2012). The disparity between the image the event creates, and the reality and cost, can sometimes be detrimental to both external and internal brandings.

From the works cited above, existing publications on urban studies mainly focus on official discourses and urban coalitions of elites, experts, and government officials rather than local voices. In other words, they focus on how the design of a mega-event fits in the urban regime and urban agenda (Burbank et al., 2001; Gunter, 2014; Maennig & Du Plessis, 2009). Other studies tend to examine how urban projects, visual displays, and propaganda help to create spectacles that represent modernity, progress, unity, and future-oriented developmental paths in the host entities. This thread of discussion focuses more on the constructed meanings of mega-events in several assigned geographical scales, as well as the social–cultural effects of mega-events.

Both internal or external branding strategies neglect how mega-events are created from a series of production, reproduction, and representation. This case study will show how Hangzhou and the Chinese state adopted both external and internal strategies to host this summit and utilise the three uses of hosting mega-event.

Hangzhou: venue city of the 2016 G20 Summit

This section introduces Hangzhou – the host city of the 2016 G20 Summit. Hangzhou is the capital city of Zhejiang Province, with an administrative land area of 16,850 km² and a total registered population of about 7.7 million (Hangzhou Statistic Bureau, 2019). Its history can be traced back to the Qin dynasty (221 BC), and its prehistorical records can be traced back to Liangzhu Culture (c. 340–2250 BC) in the Neolithic period.[1] Politically, it was the capital of the Wuyue Kingdom (AD 907–978) during the Five Dynasties and Ten Kingdoms period. It was called Lin'an, a walled city, and the temporary dynastic capital of the court of the Southern Song Dynasty (AD 1127–1279) (Fu & Cao, 2019). Despite the legitimacy of Hangzhou as a capital city in Southern Song is debatable, being a de-facto capital gave the city not only its urban layout but also its historical significance. One of the prerequisites of being a capital city, after all, was sufficient food supply. Excellent location in terms of water transportation, improved agricultural technologies, and natural fertile terrains with slopes and rivers made Hangzhou historically one of southern China's central cities in terms of food production and transportation.

As a modern city, its ability and visibility were limited internationally, not to mention it hardly competes with an influential and well-known neighbour –Shanghai. Before 2016, Hangzhou was little known internationally, and it was not even considered a "first-rate" or "first-tier "city in China.[2]

However, in domestic China, the city of Hangzhou has shown strong private sector growth since 1999, at the time the whole country was experiencing the economic restructuring of state-owned enterprises which most of them went bankrupted, sold, or closed during the late 1990s. During the same period, the city began economic restructuring and deindustrialisation, transforming it into a city with rapid growth (Qian, 2012). Moreover, the Chinese e-commerce giant Alibaba was founded in the city in this period. In the past decade, the city became a magnet for undergraduate students, entrepreneurs, and investors wishing to expand their economic territory. The municipal governments, considering the potential tax venues, proposed several policies to attract educated young people to the city. One of the indicators of urban expansion and increasing private investment of Hangzhou is the development of the new CBD, called Qianjiang New City, along the northwest shore of the Qiantang River, beginning in 2002. This new city project welcomes investments and provides financial, commercial, entertainment, institutional functions (Qian, 2011).

In November 2015, China declared that it would host the G20 Summit in 2016 in Hangzhou, with Qianjiang New City as the main venue. It was the first time China hosted the G20 Summit. For the city, hosting the high-profile G20 meeting was unprecedented and rare during the PRC government. Especially as Hangzhou had never been aligned to top Chinese venue cities, such as Beijing (hosted the 2008 Beijing Olympic Games and 2014 APEC meeting), Shanghai (hosted the 2010 World Expo), Guangzhou (hosted the 2010 Asian Games), Shenzhen (hosted the 2011 World University Games). Additionally, in August 2015, Hangzhou also applied to host the 2022 Asian Games and became the only candidate city. It received the official permission to host the games in September 2015. These efforts not only show China's ambition to host international events but also Hangzhou's rise on the stage of a host city of global events.

Presentation of the 2016 G20 Summit

G20 was initiated in 1999 and its Leaders' Summit was first held in 2008 to improve global economic governance. The 20 member countries consist of Argentina, Australia, Brazil, Canada, China, France, Germany, India, Indonesia, Italy, Japan, Mexico, Republic of Korea, Russia, Saudi Arabia, South Africa, Turkey, the United Kingdom, the United States, and the European Union (EU). Invited guests of the Leaders' Summit also include global institutions such as the United Nations, the International Monetary Fund, the World Bank, the World Trade Organization, the Financial Stability Board, the International Labour Organisation, and the Organisation for Economic Co-operation and Development (OECD) (G20OfficialWebsite, 2015).

According to Getz (2008), this summit belongs to the category of "political and state" and its content of the international meeting also fits the category of "business and trade" events. Along with the long-term developmental agenda, the 2016 G20 Hangzhou Summit was planned to take place on September 4 and 5 in the Qianjiang New City (Hangzhou's CBD) on the south bank of Qiantang River. Instead of in the crowded downtown area, the mega-event would be hosted in the developing areas of the city.

The main venue for the 2016 summit was located in the Hangzhou International Expo Center in Xiaoshan District. Thus, the underdeveloped area along the Qiangtang River and the city's former northern suburban area have become valuable and easy to develop into modernised urban sites at an accelerating pace. The construction of the centre started in 2011 and was completed in 2014. The centre was put in use in 2015 and was part of the Qianjiang New City project and not the direct product of hosting the event.

External marketing: city beautification, the G20 Blue, and the Nightscape

City beautification

The natural landscapes of Hangzhou, including the West Lake and the Qiantang River, have been well known to Chinese people. However, the appearance of the constructed environment is not that satisfying. The layout of the downtown area has been confined by its natural environment: the Qiantang River to the east, the West Lake to the west, and Wu Mountain in the south of the downtown area. In the past two decades, the city has invested in the east riverside area of the Qiantang River and the northbound area outside of the downtown area. The city has since been rapidly expanding, and the downtown area began to undergo urban renewals during 2004–2011.

The second wave of environmental improvement took place during 2015–2016 and the primary goal for this project was to improve its public image to visitors for the planned 2016 Summit. This type of project emphasises the welcoming gesture while excluding unfavourable communities in the city. Due to the short time frame, this wave of beautification focused on the short-term improvement of façades and building surfaces and the instalment of urban furniture. It concentrated on the two banks of the Qiantang River and the main entrances to the city.

There is limited information about the scope and number of the series of demolitions and renovations that took place during 2015 and 2016. But according to official news released before the summit, there were 651 infrastructure renovation projects that mainly dealt with "environmental treatment, hotels for state guests, and airport expressways" (CCTV, 2016). According to my interviews with several citizens, Hangzhou did not change its appearance on a large scale; the event simply helped the city renovate some buildings. In some areas of Hangzhou, however, residents faced large-scale demolition, undertaken for the stated purpose of improving visual integrity. For example, in the Mt. Mantou neighbourhood, more than 400 illegal buildings in the residential cluster were demolished, and 1,642 illegal rental units were emptied within approximately 100 days. As the official news report remarked: "the stationary living environment of decades is transforming at a rapid pace" (Du, 2016).

The efforts to enhance visual integrity were superficial as many façades were repainted or renovated, but the back of them remained. Thus, the real reason for the demolition was not illegality alone, but the location of the illegal building: an illegal construction in a highly visible location was a problem. Similar to most mega-events branding, this type of project emphasises the welcoming gesture while excluding unfavourable communities in the city. In the case of Rio de Janeiro during 2014 and 2016 events, the city's image has been constructed as sensational, tropical, and diverse, with its cultural heritage such as carnivals and partying, targeting European and American men (Broudehoux, 2019). These branding strategies thus trigger the demolition of buildings or removal of unfavourable styles and aesthetics.

One targeted area under severe demolition was about 20 km away from the main venue Hangzhou International Expo Center. Under the name of "Environmental Improvement Project for the G20 Summit," 5,942 housing units in 9 villages in the Qiaosi administration of Yuhang District faced demolition in the first half of 2016 alone (N/A, 2016; Tian, 2016). It is the main production centre for clothing, with 604 enterprises and more than 90% of the local population migrant workers employed in local clothing manufacturing industries. Around 2010, villagers built six-story buildings (each costing approximately 7080,000 RMB) on borrowed money, specifically for renting to migrant workers. At the time, this type of construction did not violate any regulations. Nevertheless, during the wave of demolitions five or six years later, local authorities claimed they allowed only buildings with fewer than four stories: hence these structures were illegal. Although the demolition project provided compensation to owners, the fine for illegal building per extra m^2 exceeded the compensation; each of them had to pay 100 thousand RMB beyond the amount they were compensated (Lin, 2016).

The G20 Blue, and the Nightscape

Aside from the improvements to the built-up environment, G20 Summit 2016 also requested a clear sky to advertise the achievement of environmental governance. It was not the first time China ordered the showing of a clear blue sky in such a global event. Olympics Blue (2008) and APEC Blue (2014) in Beijing have shown the world China's determination as Beijing has long suffered from air pollution, smog, and a bad reputation for its air quality. In the preparation for the G20 Summit, China took similar strategies to order pollution-prone firms, such as iron steeling firms to shut down months before the event.

The range of regulation was not limited to the Hangzhou metropolitan area, but involved hundreds of companies radiating from the centre of Hangzhou to neighbouring provinces and cities including the Yellow Mountain area in Anhui and the famous porcelain production town of Jingdezhen in Jiangxi, outward to hundred kilometres (Jin, 2016). This regulation also applied to the logistic industry, and the levels of controls were organised as the core area, the strictly controlled area, and the controlled area. Before the opening of this summit, Hangzhou was declared to be the first city in China without coal-firing, steel production, and high-emission vehicles (TVBS/FOCUS, 2016). However, this effect did not last for long after the event (Lee, 2016).

Another project, i.e. the lighting project in Qianjiang New City, the main venue of the 2016 G20 Summit, also fulfilled a sense of modernity and progress to officials and locals alike. More than 100 lighting projects were operated during this event, showing the prosperity of the city and China and its enhancement on the waters in Hangzhou. The three main areas of lighting were the West Lake, the Grand Canal, and the Qiantang River. According to statistics from the public sector, these projects earned another 35% revenue in night boat tour tickets compared with the same months from other years (Lighting, 2017).

The improvement of the urban landscape, air quality, and lighting have changed the visual experience of the city. However, this kind of make-up of the city did not bring many positive impacts to ordinary people. For them, these efforts put forward for the event did not improve their daily lives but instead saved face for the state. These kinds of beautifying projects were known as "face projects", meaning the motives behind them are for reputation, and the beautification practice is shallow and of surface value. However, for those who are pro-growth, the project has been positive. One official said that the lighting projects for the G20 summit have greatly improved the appearance of Hangzhou and has caused it to be considered a first-rate city. The official added, "since then, we don't have to introduce [Hangzhou] as the city 150 kilometres away from Shanghai" (Lighting, 2017).

The beautification projects for the event eventually showed off the city's man-made beauty and how it emphasises the natural landscape of the city. Though the target audience group for these projects was not the citizens who had paid the price to adapt to this change, the general plans of improvement and beautification, have since been accepted by the host city.

Internal marketing: lessons on hospitality and passive docility

Citizens engagement

Physical improvement was the first step to establish the confidence of being a modern and prospering host city. However, to reach the goals for physical improvement on such a large geographical scale, the host city required the coop-eration of local citizens and used different strategies to create new social norms for local people.

The first strategy was to awaken nationalist sentiment in citizens to share the pride and prosperity of being the host city for the nation. This strategy, though, was not undertaken by advocating patriotism directly in propaganda. The official slogan of this event was "A good host, a better G20". The rationales of being "good" hosts came from two concerns: security and the "face" –to assure the safety and the enjoyment of the guests. Therefore, the Hangzhou people were cooperative with many arrangements for this event. According to an official sur-vey, 96.8% of Hangzhou residents showed their support and agreement of this event, and this event involved 760,000 volunteers. More than 26,000 school and college students applied to be volunteers in the G20 Summit with 6,000 chosen after the first round of selection (Boya Tan, 2016)

Local authorities and news channels focused on these good virtues inherited by the people in Hangzhou and then promoted them through several campaigns and programmes in the name of hospitality- an even softer appeal than nationalism or patriotism. Enthusiastic volunteers joined programmes to support the progress of the events. For example, programmes aimed for this event, such as English learn-ing, training for taxi drivers, and volunteer programmes for retired citizens and college students, gave opportunities for locals to be part of the event. These pro-grammes also included lessons for introducing local traditions, religious practices, tourist attractions, and Hangzhou cuisine. In other words, they were designed not only to support the event and the guests but also to teach local people how to brand the city to others. According to two student volunteers from Zhejiang University, the training courses included English conversational skills, Chinese culture, and inside information about the event and the venues. Students were eager to be part of the event because it gave them the rare chance to see global leaders in the flesh.

Forced tourism

Despite the event requesting local participation and welcoming gestures to foreign guests, citizens were not welcome in both the preparation and the event itself. One month before the event, the city was full of checkpoints in every public space, and hotels were being checked into as a way to prevent protests or illegal gather-ings. September 1–7 became days-off for people working in urban Hangzhou, including nine districts. Moreover, during the period of the summit, citizens were encouraged to leave the city. A quarter of citizens were forced to go for tourist breaks and, with their IDs, could travel to neighbouring provinces and scenic areas at discounted prices. Migrant workers went back home since many of their employers shut down for the period. Dissidents and potential troublemakers on record were "invited" to suburban areas with police to participate in rural tourism under supervision.

It was ironic that residents in this famous tourist city had to leave their homes in order to greet the guests. However, when I conducted face-to-face interviews with the Hangzhou people about the inconvenience which resulted from this event, the seemingly docile Hangzhou people admitted to this but added that these regulations, and the inconvenience, were quite necessary to minimise the risks "for the state's face" like pollution, deteriorating buildings, social protests, and insecurity caused negative impression or physical hazard such as fire or traffics. Some of them even felt proud to be part of the international event, despite these controls.

Netizens[3] and the muted cyberspace

Another reason for this tolerant attitude is the censorship of speech. As the event approached, online speech underwent stricter supervision than usual and reached a peak on the first day of the summit (Allen-Ebrahimian, 2016). Some internet administrators closed online forums temporarily to prevent them from being hacked. In July 2016, Enping Kuo, a public officer working at a street office in Zhejiang, wrote a post titled: "Hangzhou, shame on you" on his tweet-like blog. In this post, he mentioned that the expense from the public sector was "astronomical" that accounted for 70% of the city's annual revenue. One of his points was that Hangzhou had been over-preparing and overspending on this event due to a lack of confidence, and these efforts in turn stopped visitors from seeing the real beauty of Hangzhou, which is a delicate city in nature. He also criticised the disturbance caused by this event, "if the disturbance of everyday life is the expected cost of such an international event… then it is doomed to fail before it starts". This post was deleted shortly afterwards, and Kuo was arrested for 10 days under the name of web information abuse and disobedience of social order. He became unemployed afterwards.

This case further quietened the public "netizens" of China, who have been under strict censorship already. Studies of Chinese netizens, emerging internet citizens, show how the "new opinion class" is influencing the state-society relation of contemporary China (Qiang, 2011). For Chinese citizens, there had been only official sources of information from mass media, and the lack of public debates and the strict censorship provided little room for public speech. As an experienced host state, China has been shutting down online forums during mega-events as in the 2008 Beijing Olympics Games (Qiang, 2011).

Same in the case of G20 Hangzhou Summit, local online forums and that of local universities were shut down during the G20 Summit days for unexplained reasons. Hangzhou netizens, emerging internet citizens, expressed their objections and disagreement online, and cyberspaces became areas where isolated audience members and netizens could discuss and debate the event. From the voices of Hangzhou netizens, other netizens could understand what was happening in the frenzied hosting city. In the case of Rio, the activists against the pro-event coalition used pamphlet or mural to denounce the event and its supports (Broudehoux, 2019). In Hangzhou, there was no reported case of any physical gathering but also a slogan invented to ridicule the official one: "An expelled host, a better G20". Since a collectively and liberally public realm, whether physical or virtual, is not foreseeable, random speech by netizens became not only the voice but also the indicator to see the boundary between acceptable public speech and something in violation of the laws. Local authorities set up a website called "G20 rumour breaking forum" in August 2016 to refute rumours about this summit. This apparatus was a strategy to refute those "counter-discourse" which netizens were spreading and eventually the event came to an end without any counter-voice heard by the visitors.

In sum, the summit worked as a way for place-event marketing in which the branding strategies, whether external or internal, followed the priority to satisfy the needs of the urban products' consumers. The exclusion of migrant workers, people living in the urban fringe, active netizens at the time for preparing the summit not only demonstrated how a city expected to be represented at a mega-event but a new social order in formation. The setting up of this site could be viewed to stabilise social order, as the Hangzhou people were anxious about the changes, rules, and adjustments required for this unprecedented event.

Ultimate branding of the spectacular city and the lake

Despite of all these debates and disagreements, the event turned out to be a success and secured the public face of China. What dissolved the disparities of image-making and social controls of the city was the spectacular West Lake, "the face of the city," as cultural leaders in Hangzhou recognised. A month before the event, the city was under the strictest supervision of security checks. As the event approached, fewer people were staying in Hangzhou. The West Lake, the most popular site throughout the year in Hangzhou, became emptied and the scenic area was closed from August 20 to September 6. Though the lake and the surrounding area were not used for meetings, several hotels around the West Lake accommodated global leaders. For people living within the area, local authorities outlawed cooking with a gas stove around the West Lake (Cheng, 2016) and asked them to keep all windows closed, providing a subsidy for the extra cost of air conditioning.

West Lake was part of the spectacle of the opening concert. The theme of the concert was "Hangzhou, A Living Poem." Zhang Yimou was the chief director of the G20 show, and he also directed the opening ceremonies of the 2008 Beijing Olympic Games and the 2014 APEC Reception in Beijing. Though the performance and its related fireworks were not open to the public, these were broadcast on live TV channels and websites from the single source of China Central Television (CCTV), the official media channel. Relevant news reports on this summit garnered a 5.36% TV rating among all people watching TV at the same time, the highest over other news channels in 2016 China, and the concert 26.49% (Sun, 2016).

In contrast to the tight social controls and the closure of most public areas leading up to the event, the outdoor opening gala was a spectacle welcoming to most Chinese and Hangzhou people, even those who were being directly prevented from approaching West Lake at that time. The platform was set in the lake, 3 millimetres below the water's surface; the design made the lake itself the background and the stage where performers and dancers moved with the splash of water. The one-hour show included nine programs in the integrated Chinese and Western forms of art: symphony orchestra, ballet dance, folk dance, Yue Opera, vocals, poetry reading, Chinese zither, and cello ensemble, and piano solo.

This show was designed not only for the international audience but the domestic audience as well. Most of the locals were unable to attend in person, nor were they able to walk the streets freely during those two days, but they watched real-time live video platforms of the concert as a major way to feel involved in this event. To collect the reaction to this performance, I collected information and comments from the bullet-screen of this video and analysed how Chinese netizens, as a collective, responded to this show. Surprisingly, this show provided not only a sense of pride, but it also helped ease months of social tension between the Hangzhou residents and the authorities. These real-time comments, or "bullet-screens" as called by the Chinese netizens, were shown in random order alongside the progress of the show.

From the information on the bullet-screen collected on the platform "LE TV", these performances conveyed the message of cultural superiority and economic power –that this beautiful performance could happen in a city other than Beijing was refreshing to a Chinese audience. Some netizens described the scene as referencing ancient forms of tribute and China's traditional role as the perceiver. The most conflict-prone comments related to the expenditure of this event and the display of wealth. Simply put, the comments asked whether all of these efforts, regulations, and the concert itself, were designed solely to "save the face" of the nation. Despite these small conflicting comments on the bullet-screen, the debates between event expense, economic power, cultural superiority, local pride, and nationalism were distilled into a harmonic rosy picture about the past and the future of the city. Netizens expressed excitement about the foreseeable prosperous boom in tourism in Hangzhou and praised the city for its naturally beautiful landscape and the burgeoning tourism. In other words, the visual spectacle smoothened the potential power of counter-discourses. Most importantly, cyberspace, despite the censorship, was still the only public realm where isolated audience members and netizens could discuss and debate the event.

As a result, the event went smoothly, and there was no social protest at this event. Netizens appearing on the live-streamed program showed pride toward the city or/and the nation. Whether local or not, citizens felt proud. For the Hangzhou residents, the city was getting global visibility, and its cultural traditions and beautiful landscape were considered superior to other first-rate cities. For non-Hangzhou residents, they appreciated the city in this event, and they felt proud to be part of the Chinese nation. For those who had not previously recognised this rising city, they felt proud as well. They may still have regarded Hangzhou a second-tiered city, but one powerful enough to host a global event and represent another soft side to China, distinct from the political one of Beijing, to the world. Besides the abstract proud emotion, Hangzhou and its tourism product became more popular. After the event was over, the managing company opened the meeting halls for tourism since September 25 as "G20 Experience Venue". According to Ctrip tourism agency, the largest one in China, online searches for tourism in

Hangzhou went doubled right after the event. Several weeks after the end of the event, the sales of travel-related service had increased more than 40% than the same time in 2015 (Yujie Wang, 2018).

Conclusion: the venue city in the making

For Hangzhou, this event was a promotional strategy both internationally and domestically. Hangzhou is seen as a city guaranteed by the Chinese Communist Party and the Chinese state as an essential base of commerce and global events. After the G20 Summit ended, the Hangzhou municipal party committee and municipal government addressed to its citizens a thank-you note on September 7, 2016, for their cooperation during the event. This note was released on *Hangzhou Daily*, the official newspaper and its web site, the next morning (http: //hangzhou.zjol.com.cn/system/2016/09/09/021293453.shtml). It expressed gratitude to citizens for being good hosts to visitors and concluded with the following words:

> The post-summit era of Hangzhou has come: the fruits of the summit have translated into a sense of achievement for our citizens; the summit experience on social governance is as fulfilling as a long-term management mechanism; the success of the summit will become a new start for Hangzhou towards internationalization… .

This passage emphasises official rhetoric regarding the spirit of the city rather than the mere physical betterment. Also, it indicates that the success of the event is only the beginning for Hangzhou, which implies the coming of the next event i.e. the Asian Games in 2022. As this chapter demonstrates, the city used strategies to improve the perception of pollution, the facade of a built-up environment, and the strengthened control of logistics and transportation as part of external branding strategies. Besides, Hangzhou authorities also implemented soft strategies such as social education and volunteer programs to enhance local identities and the responsibility of being part of the Chinese nation. As studies have shown, the disparities between social control during mega-events and everyday realities are often sources of social unrest. The spectacle on the lakeshore soothed disagreements and consolidated local identity in citizens, which is arguably one of the most important, but invisible, event legacies of the G20 Summit in Hangzhou.

The case study of the 2016 G20 Summit in Hangzhou demonstrates how a host city in a developing country can utilise both external branding and internal branding to reach its goal as a competitive city and a future venue city, which will be an upgraded form from a traditional tourism city. From the aspect of external branding, Hangzhou's case as the host city had triggered demolition on the urban–rural fringe and a wave of beautification of the downtown area. These efforts echoed with its ongoing agenda to urban expansion and developing riverside areas along the Qiantang River. Lighting projects and regulations on production also helped create a better visual impression to the visitors.

As for internal branding, this event has enhanced local pride and identity as citizens were proud of the city's performance and overall improvement. The disputes over demolition and tight controls over society and cyberspace were muted by local states and the city was represented as a welcoming place-brand for tourism and investment. This event helped the city reach its peak concerning its competitiveness to other cities in China because of the improvements to its environment, the decreased pollution, and economic development caused by the event organisation. However, a new social order introduced by the authorities, mixed with strict controls over industries, tourism, logistics, transportation, and communication temporarily made the city a contested realm and a visual ghost town. These disadvantages impacted the residents and influenced their perception of the event and its marketing outcomes.

Notes

1 From UNESCO World Heritage Centre (2019).
2 The Chinese city tiered system uses GDP, administrative levels, and population as three main criteria to categorise its cities into four clusters: the first-tier cities, second-tier cities, third-tier cities, and fourth-tier cities.
3 A netizen is a user of the Internet, often one who is actively engaged in uncensored online discussions of political and social issues (Netizen, 2020).

References

Allen-Ebrahimian, B. (2016). China's G-20: The most censored day of the year. *Foreign Policy*. Retrieved from https://foreignpolicy.com/2016/09/06/chinas-g20-the-most-cen sored-time-of-the-year-hangzhou-weibo-empty-streets/

Anholt, S. (2008, 01 Feb.). Place branding: Is it marketing, or isn't it? *Place Branding and Public Diplomacy, 4*(1), 1–6. https://doi.org/10.1057/palgrave.pb.6000088

Arnegger, J., & Herz, M. (2016). Economic and destination image impacts of mega-events in emerging tourist destinations. *Journal of Destination Marketing & Management, 5*(2), 76–85.

Ashworth, G., & Kavaratzis, M. (2009). Beyond the logo: Brand management for cities. *Journal of Brand Management, 16*(8), 520–531.

Boya Tan, M.Y. (2016). Millions of volunteers in Hangzhou aim to help with the G20 Summit/杭州百万志愿者服务G20. *China Youth Daily*. http://world.people.com.cn/n1/2016/0803/c1002-28607856.html

Braathen, E., Mascarenhas, G., & Sørbøe, C.M. (2016). A 'city of exception'? Rio de Janeiro and the disputed social legacy of the 2014 and 2016 sports mega-events. In V. Viehoff & G. Poynter (Eds.), *Mega-event Cities: Urban Legacies of Global Sports Events*. Ashgate.

Broudehoux, A.-M. (2007). Spectacular Beijing: the conspicuous construction of an Olympic metropolis. *Journal of Urban Affairs, 29*(4), 383–399.

Broudehoux, A.-M. (2019). Resisting Rio de Janeiro's event-led place promotion: From insurgent rebranding to festive counter-spectacle. In *Urban Events, Place Branding and Promotion* (pp. 124–140). Routledge.

Burbank, M., Andranovich, G., & Heying, C.H. (2001). *Olympic Dreams: The Impact of Mega-events on Local Politics*. Lynne Rienner Publishers.

CCTV. (2016, 28 Aug.). Hangzhou ready to host world leaders. http://english.cctv.com/2 016/08/28/VIDEijaxsgB4LBpLCtyz0P7c160828.shtml

Cheng, X. (2016, 27 Aug.). G20 is Coming, for whom the Sky in Hangzhou Becomes Blue?/ G20將至杭州的天空為誰而藍? *Epoch Times*. Retrieved from https://www.epo chtimes.com/gb/16/8/27/n8242037.htm

Cope, B. (2015). Euro 2012 in Poland: Recalibrations of statehood in Eastern Europe. *European Urban and Regional Studies*, *22*(2), 161–175.

Cudny, W. (2016). *Festivalisation of Urban Spaces: Factors, Processes and Effects*. Springer.

Cudny, W. (2020). The concept of place event marketing. Setting the agenda. In W. Cudny (Ed.), *Urban Events, Place Branding and Promotion: Place Event Marketing*, pp. 1–24, Routledge.

Fu, C., & Cao, W. (2019). *Introduction to the Urban History of China*. Springer.

G20OfficialWebsite. (2015). About G20. http://www.g20chn.org/English/aboutg20/A boutG20/201511/t20151127_1609.html

García, B. (2004). Urban regenration, arts programming and major events. *International Journal of Cultural Policy*, *10*(1), 103–118. https://doi.org/10.1080/10286630420 00212355

Getz, D. (2007). *Event Studies: Theory, Research and Policy for Planned Events*. Elsevier.

Getz, D. (2008). Event tourism: Definition, evolution, and research. *Tourism Management*, *29*(3), 403–428.

Gunter, A. (2014). Mega events as a pretext for infrastructural development: The case of the All African Games Athletes Village, Alexandra, Johannesburg. *Bulletin of Geography. Socio-Economic Series*, *23*(23), 39–52.

Hangzhou Statistic Bureau. (2019). *2019 Year Report*. Hangzhou, Zhejiang Retrieved from http://tjj.hangzhou.gov.cn/art/2019/10/23/art_1653175_39222973.html

Jin, J. (2016). *On G20 Hangzhou Summit*. https://www.chinesepen.org/blog/archives/69209

Joo, Y.-M., Bae, Y., & Kassens-Noor, E. (2017). Mega-events and mega-ambitions: South Korea's rise and the strategic use of the big four events. In *Mega-Events and Mega-Ambitions: South Korea's Rise and the Strategic Use of the Big Four Events* (pp. 1–22). Springer.

Kavaratzis, M., & Ashworth, G.J. (2005). City branding: An effective assertion of identity or a transitory marketing trick? *Tijdschrift voor economische en sociale geografie*, *96*(5), 506–514.

Knott, B., Fyall, A., & Jones, I. (2015). The nation branding opportunities provided by a sport mega-event: South Africa and the 2010 FIFA World Cup. *Journal of Destination Marketing & Management*, *4*(1), 46–56. https://doi.org/10.1016/j.jdmm.2014.09.001

Lee, C.-Y. (2016). G20 in China: effective or irrelevant? *RSIS Commentary*. https://www .rsis.edu.sg/rsis-publication/cms/co16232-g20-in-china-effective-or-irrelevant/

Lenskyj, H.J. (2008). *Olympic Industry Resistance: Challenging Olympic Power and Propaganda*. State University of New York Press.

Lighting, C. (2017). Lights in G20 transformed Hangzhou into a first-tier city/ G20灯光让杭州瞬间变成了一线城市. *City Lighting*, *40*, 14.

Lin, R. (2016, 9 Apr.). Hangzhou forcing demolition to welcome the G20 summit hundreds of police arrested people at the midnight /杭州強拆迎G20峰會 數百警察 半夜突襲抓人. *The Epoch Times*. http://www.epochtimes.com/b5/8/12/30/n2379493 .htm

Maennig, W., & Du Plessis, S. (2009). Sport stadia, sporting events and urban development: International experience and the ambitions of Durban. *Urban Forum*, *20*, 61–76.

Mistilis, N., & Dwyer, L. (2000). Information technology and service standards in MICE tourism. *Journal of Convention & Exhibition Management*, 2(1), 55–65. https://doi.org/10.1300/J143v02n01_04

N/A. (2016, 6 May). Yuhang Qiaosi Cadres led to a demolition of thousands illegal building in 21 days/ 餘杭喬司街道黨員幹部帶頭21天拆了千餘戶違建房. *kknews*. http://www.epochtimes.com/b5/8/12/30/n2379493.htm

Netizen (2020). *Dictionary.com*. Retrieved 17 June 2020 from https://www.dictionary.com/browse/netizen

Ooi, C.-S. (2010). *Branding Cities, Changing Societies*. Copenhagen Business School.

Qian, Z. (2011). Building Hangzhou's new city center: Mega project development and entrepreneurial urban governance in China. *Asian Geographer*, 28(1), 3–19. https://doi.org/10.1080/10225706.2011.577977

Qian, Z. (2012). Post-reform urban restructuring in China: The case of Hangzhou 1990–2010. *Town Planning Review*, 83(4), 431–456.

Qiang, X. (2011). Liberation technology: The battle for the Chinese internet. *Journal of Democracy*, 22(2), 47–61.

Richards, G. (2015). Developing the eventful city: Time, space and urban identity. *Planning for Event Cities*, 37–46.

Richmond, M. (2016). The urban impacts of Rio's Mega-events: The view from tow 'unspectacular' favelas. In V. Viehoff & G. Poynter (Eds.), *Mega-event Cities: Urban Legacies of Global Sports Events*. Ashgate.

Shin, H.B., & Li, B. (2013). Whose games? The costs of being "Olympic citizens" in Beijing. *Environment and Urbanization*, 25(2), 559–576.

Smith, A. (2012). *Events and Urban Regeneration: The Strategic Use of Events to Revitalise Cities*. Routledge.

Sun, X. (2016). Those usage data that you don't know about CCTV/ 那些你所不知道的央視數據. http://1118.cctv.com/2016/09/30/ARTIJxsxspaTb5tpEiaeILGr160930.shtml

Tian, X. (2016, 02 Jul.). Demolition in G20 Hangzhou summit: Here comes red guards again/G20峰會杭州拆遷: 紅衛兵又來了. *Banned Book*.

Timms, J. (2012). The Olympics as a platform for protest: A case study of the London 2012 'ethical' games and the play fair campaign for workers' rights. *Leisure Studies*, 31(3), 355–372.

TVBS/FOCUS. (2016, 17 Aug.). Hangzhou blue, or G20 blue, a necessity for a globcal summit / 「杭州藍」又稱G20藍 全球峰會必先備. *TVBS/FOCUS*. https://news.tvbs.com.tw/world/669413

UNESCO World Heritage Centre (2019). *Archaeological Ruins of Liangzhu City*. Retrived 17 June 2020 from https://whc.unesco.org/en/list/1592

Yujie Wang, X.Z., Xin Zhong, & Sijia Lv. (2018). Research on the influence and promotion path of major festival activities on urban tourism internationalization: Taking Hangzhou G20 summit as an example/ 重大节事活动对城市旅游国际化的影响与提升路径研究——以杭州G20峰会为例. *World Economic Research*, 7(4), 125–133.

Zhang, L., & Zhao, S.X. (2009). City branding and the Olympic effect: A case study of Beijing. *Cities*, 26(5), 245–254.

http://hangzhou.zjol.com.cn/system/2016/09/09/021293453.shtml

3 Branding Shanghai as a global city through China Shanghai International Arts Festival

Yifan Xu

Introduction

This chapter encompasses the research topics of festivals and place branding. There is a large portfolio of publications devoted to festivals and events, and the issue of place branding is widely studied. Festivals and events have important impacts on host areas. They influence society, culture, economy, and employment in destinations where festivals are organised. The impact of festivals is especially visible in cities, where it is often referred to as the festivalisation of urban spaces (Cudny 2016). One of the important results of a festival's organisation is much neglected in science, that is, its influence on image and destination brand (Lee and Arcodia 2011).

Studies of festivals' relation with city branding, however, are very rare. Thus, there are not many comprehensive scientific contributions devoted to the role of events and festivals in the branding of cities (Garcia 2004; Broudehoux 2017; Cudny 2019). To fill in the research gap this chapter investigates a case study of government-funded large-scale festivals with core programmes in performing arts, the China Shanghai International Arts Festival (henceforth referred to as CSIAF). The CSIAF was selected as a case study to show how large-scale cultural events influence the creation of a place brand. Specifically, this chapter speaks to these gaps in the fields of festival studies and city branding by examining the potential of the CSIAF in branding the city of Shanghai.

The festival brings additional complexity in the analysis of its branding mechanisms due to the symbolic meanings embedded in cultural products at different times and spaces. The case also adds to the empirical research on city branding practices in the Asia-Pacific region. This chapter focuses on the place branding approaches and messages of the most recent CSIAF in 2018 as it was the 20th anniversary of the festival and the 40th anniversary of the "Reform and Opening-up" in China. The main aim of the chapter is hence to analyse the CSIAF's contribution to place brand creation; therefore, the research question encompasses how the festival brands the city of Shanghai, and, in some instances, the nation of China.

The chapter is divided into ten sections. The theoretical background follows the introduction, where the relations between festivals and place branding with a focus on urban spaces are under scrutiny. The third part presents the qualitative methods applied throughout the chapter. The following two parts of the chapter present a brief history and cultural development of Shanghai as the host city of the CSIAF, as well as the festival's organisation, history, mission, stakeholders, and programmes. Following a general discussion of how the festival exerts influences on the image-making agendas of Shanghai and China, the sixth part introduces the 2018 CSIAF as a more comprehensive place brand product created by the city and the state. The seventh to ninth parts of this chapter discuss the similarities and differences in the themes of city branding and nation branding towards domestic and international audiences through the 2018 CSIAF. This chapter ends with conclusions presenting the most important findings on the relationship between the festival and its branding effects on Shanghai and China.

Theoretical background

International arts festivals

Festivals are among the most important category of planned events (Getz 2008). According to Getz's (2008) categorisation of events, it is important to define festivals in contrast to other events types (e.g. political, business, educational, or sports events). Falassi (1987,) defined festivals as

> a periodically recurrent social occasion in which, through a municipality of forms and a series of coordinated events, participate directly or indirectly and to various degrees, all members of the whole community, united by ethnic, linguistic, religious, historical bonds and sharing a worldview. (p.2)

Recent years witnessed the diversification of festival scales, genres, themes, geographical locations, and impacts (Quinn 2005; Klaic 2014; Gold 2016; Wilson et al. 2017). Festivals can be devoted to culture, some to science (science festivals) or food and wine, others are multicultural, etc. (Cudny 2016). Arts festivals are a subset of festivals that feature arts and cultural programs. Events such as theatres and art exhibitions are organised in arts festivals to create a celebrative atmosphere for their targeting audiences, oftentimes local communities, and tourists. As the globalisation process accelerated, arts festivals have gradually introduced international elements, through diversified programs and genres, and themes relating to international issues (Gold 2016). Recent decades saw the boom of international arts festivals across different continents, especially in the new Millennium (Foley et al. 2012; Dong 2015).

Festivals exert many tangible and intangible impacts on different scales of places (Cudny 2016). These impacts are often studied as results coming from tourism induced by festivals i.e. festival tourism (Getz 1991; Cudny 2013). This type of tourism influences local, regional, and sometimes state economies (Kim and Uysal 2003). However, festivals' impacts cannot be limited to economic effects only. Festivals have long been well researched as a social and cultural phenomenon in anthropology and sociology regarding the attendees' experience and the meaning-making processes within local communities (Getz 2008, 2010; Klaic

2014). Festivals also add to the local cultural atmosphere and improve cultural facilities that are attractive to external investment and creative class (Bianchini and Parkinson 1994; Florida 2012).

Festivals are hence multi-dimensional phenomena (Getz 2010; Klett 2017), influencing local art and culture (Quinn 2005), social relations (Arcodia and Whitford 2007), and the image and brand of host destinations (Cudny 2016). Richards and Palmer (2010) argued that eventful cities are critical to the sustainable revitalisation of post-industrial cities, with regard to their physical, economic, social, and cultural vitalities. As such, festivals are often part of placemaking strategies or so-called event-led regeneration programs (Garcia 2009; Getz 2010; Dwyer and Beavers 2011; Sacco et al. 2014).

Place branding through festivals

Place product development, image creation, and promotion are intertwined parts of branding endeavors (Cudny 2019). Based on neo-liberal marketing rules, a place like a city could be treated as a product or a business (Ashworth and Voogd 1994; Kavaratzis 2004) that offers a variety of products (like land, services, investment opportunities, tourist attractions, culture, and events) (Hanna and Rowley 2008; Cudny 2019) and therefore could be marketed, branded, and sold to different groups of clients (e.g. inhabitants, immigrants, tourists, entrepreneurs). Many authors defined city branding as a complex socio-economic urban development strategy aimed at the creation of a city brand attractive for inhabitants, tourists, and investors (entrepreneurs) (Ward 2005; Kavaratzis and Ashworth 2005; Ashworth and Kavaratzis 2009; Anholt 2008, 2010; Zenker and Beckmann 2013; Cudny 2019).

Accordingly, city branding is subject to a long-term development plan of place product rather than a simple marketing communication strategy. However, proper place branding could not exist without a promotional element. Many scholars have recognised the criticality of city "image formulation and communication" (Kavaratiz 2004, p. 62) in branding a city. Ruzinskaite (2015) argued that "place image is delivered through a variety of transmission channels to receivers who decode it" (p. 90). Hence an effective city branding approach should focus not only on the explicit messages communicated to audiences but also on other types of interventions, such as culture and overall design (Kampschulte 1999).

With its ready-made arts and cultural programs as branding assets, urban arts festivals have been valued as "a sort of 'quick fix' solution to their image problems" (Quinn 2005, p. 932). Festivals and cultural events are often regarded as displays, spectacles, or symbols of local cultural and political agendas (Ashworth 2009; Karvelyte 2017), contributing to the overall city branding mix. They provide the city brands not only a symbol for the city to be associated with but "an impression of a memorable journey to a given city, generating pleasant experiences and memories connected with it in order to persuade the recipient of the marketing message" (Cudny 2016, p. 88).

Cudny (2016) suggested that the production of a city's image space is through the interaction between subjective and objective urban spaces. This framework highlights the experiential aspect of city image-making, which is concerned with people's subjective perceptions through experiencing the social and cultural atmosphere in a city and the physical environment, infrastructure, and landmarks. Together they provide a comprehensive experience for both visitors and residents to construct perceptions of the city image out of their interactions with the cultural offerings in the city.

Taking the perspective of festival tourism and destination marketing, Cudny (2016) suggested that an important function of cultural events is not only "creating the image of host cities" but making "an attractive place product" (p. 88). According to Cudny (2019), the branding role of events and festivals in relation to urban places may be referred to as place event marketing. It is understood as "all activities aimed at enriching the product of a given city by offering its consumers a well-chosen, interesting and diverse portfolio of events" (p. 16).

There are two important investigating perspectives from the aforementioned discussion of place branding through festivals, corresponding to the key aspects in place event marketing (Cudny 2019). The use of events including festivals is first and foremost aimed at developing a product of city brands. The ultimate goal of festivals is to create unique and memorable experiences through diverse programs on-site. Festivals are city products purposefully constructed for different audiences, based on the extraordinary experiences that people are searching for. Such experiences (like music, dance, theatre, and film) are part of festival programs that provide a unique festival atmosphere and participation possibilities (Kavaratiz 2004).

A city is also promoted through events. As festivals attract media attention (due to celebrities' presence, interesting programs, mass tourism, etc.), they are widely presented both in traditional (TV and press) and electronic (Internet) media. Cities could gain an adequate amount of media exposure as the host of or partners with festivals, or simply as the background of events. Such media exposure hence contributes to the city's image-making agendas (Ooi and Pedersen 2010; Cudny 2016; Kavelyte 2017). Consequently, different communication channels are used for these media exposure of festivals to facilitate place branding, through sending place image-related messages to the targeted audiences.

Since place branding is a multi-dimensional policy combining the interests of public institutions, private entities, local cultural sectors, politicians, etc. (Lucarelli 2018), place image is collectively created. In organising festivals, these stakeholders establish different relationships with festival organisers, who might be more interested in promoting the festival instead of the place (Waterman 1998; Todd et al. 2017). Therefore, the effects of place branding through festivals are uncertain (Hildreth 2010). Additionally, it must be stressed that city branding is not only aimed at external recipients (like tourists immigrants, and entrepreneurs) but also at internal recipients (i.e. city inhabitants) (Kavaratzis and Ashworth 2005; Karvelyte 2017). Cudny (2019) also recognised that large-scale events and festivals are one of the major place branding tools that apply not only to cities but nations.

Methodology

This chapter applied qualitative methods, allowing an in-depth case study of the China Shanghai International Arts Festival (Crang 2003; Yin 2017). Analysis of such secondary sources as reports, documents, media, and internet resources are important qualitative research methods used in social studies (Riffe et al. 2014). Among the most important methods used in this chapter are literature analysis, internet search, and document analysis.

Literature analysis was used to find themes, theories, and case studies presented by other authors, related to the concepts highlighted in this case study (Cudny 2019). The literature analysis was conducted with the use of the following databases: China National Knowledge Infrastructure (CNKI), Web of Science, and GoogleScholar. This analysis encompassed such concepts as festivals, place and city branding, and festival impacts.

The internet analysis used the keyword search of media presentations of the 2018 China Shanghai International Arts Festival (the 2018 CSIAF) in English and Chinese, on Google.com and Baidu.com (the Chinese version of Google). As a result, 51 news reports and commentaries related to the festival were extracted from Baidu and 13 from Google during Summer 2019. Promotional videos of the 2018 CSIAF were extracted from YouTube and Tencent Video (tenent.qq.com, the CSIAF's media partner in 2018) as other common communication channels for place branding through festivals, in addition to news articles from the official website of the CSIAF.

In this chapter, documents and internet analysis allowed obtaining information about the festival i.e. its organisational structure, aims, program, etc., as well as the branding efforts of the city of Shanghai and China. Moreover, the analysis of the media and the internet was a source of information regarding the promotional and branding impact of the festival through messages framed in the media (Cudny 2019). Content and thematic analyses were applied to these secondary data, such as policy documents and literature for the comparison and synthesis of themes and data triangulation (Mayring 2004; Thomas and Harden 2008).

This research considered both the festival's experiential and symbolic associations with the place brands in Shanghai and China (Ruzinskaite 2015), particularly through the cultural and social encounters at festivals and the symbolic meanings embedded in different media communication channels. It used festival programs, the physical festival atmosphere, and a small selection of interviews with festival attendees as the proxy for the festival experience created collectively by festival organisers and their stakeholders and non-stakeholders (Wilson et al. 2017).

The secondary sources of data included photo records, videos, TV broadcasts, news reports, policy documents, festival programs, sponsors' websites, government websites, and also the fetival's website. Using the qualitative data analysis software Nvivo, thematic analysis was conducted identifying different messages conveyed to festival audiences at home and abroad, in relation to the city and nation brand agendas. Content analysis of festival programs was conducted to uncover the symbolic meanings of cultural display and event arrangements.

Shanghai as the festival host city

Location and history of Shanghai

The geographical location of Shanghai grants the city advantages over other coastal cities. Located on the coast of the East China Sea at the intersection of Yangtze River, Huangpu River, and Suzhou River, the city historically served as a critical port for foreign shipment to come to China and Chinese goods going out. Originally an administrative unit for trading and exchanges in the Jiangnan region (the southern Yangzte River Delta area), Shanghai has been critical in facilitating the economic development of the region for centuries. In the meantime, influenced by *Jiangnan culture*, as well as domestic and foreign cultures through business and trade, the city has developed an open mind to different cultural values.

Growing out of a market town in the thirteenth century, Shanghai had experienced drastic stages of changes over its history of urbanisation (Wong 2014). The Treaty of Nanjing built the foundation for Shanghai to become the modern city of China in the early twentieth century, as "an important treaty port and financial center in Asia" (p. 99). The city's functional positioning was drastically changed from an outward-looking, consumption-driven, and comparatively independent global city in the 1920s to a city that is conservative, production-based, and responsive to national agendas in the socialist reform period (1949–1977) of the People's Republic of China. The city's economic development was hence downplayed but later re-prioritised in China's Reform and Opening-Up era.

It is not until 1992 when Shanghai's economy witnessed a significant surge, due to the establishment of the Pudong District. The city has since developed its cultural infrastructure, restructured its economy, and hosted many international events, such as the 2010 Shanghai EXPO. With its distance away from Beijing – China's political center, its increasing average GDP, and a well-established cultural exchange environment and connections due to the golden era of the 1920s, Shanghai is regarded as a preferable place for large events. It obtains an advantageous set of cultural, economic, and political assets that are necessary for successful event planning and implementation, in comparison to other Chinese cities.

With an urban area of 2,643 square kilometers and the total land area of 6,219 square kilometers, Shanghai is one of China's mega-city.is the most populous city[1] in China, with a population of 26.32 million in 2019.[2] The city's population growth since the 1980s could be attributed to the internal migration of workers from surrounding provinces, such as Anhui, Jiangsu, and Zhejiang. Amongst the majority of Han Chinese ethnicity are a growing number of minority ethnicities and over 150,000 foreigners who were officially registered in the 2010 census.[3] The city also witnessed an expansion of its administrative area during the socialist reform era, when suburban districts had been gradually added to Shanghai's metropolitan area. Downtown Shanghai includes districts which were concessions for foreigners in the 1920s, clustered along the west bank of Suzhou River.

The Reform and Opening-Up era revitalised Shanghai's geographical advantages as a port to reconnect China with the world through trade and business services, and to increase its status nation-wide as one of the fastest developed coastal cities in China. The city's GDP doubled from 2009, regardless of the global economic recession. Average residents in Shanghai had more disposable income in 2017 than in any other cities in mainland China, reaching about 59,000 yuan (9,316 U.S. dollars). With its growing economic independence, Shanghai has obtained more freedom in experimenting with local initiatives. Now the city pioneers in the creation of a socialist modern city model in China. Also, Shanghai had recognised the economic value of integrating the city into the Yangtze River Delta region by taking initiatives to lead the comprehensive and collective regional economic development, through resource sharing, market development, and trade activities.

Shanghai's cultural sector

Contemporary Shanghai's cultural sector is characterised by the commercialisation of culture. It is attributable to Shanghai's well-established cultural market and the entrepreneurial policy agendas (Wu 2004; Lin 2018). In the early stage of Reform and Opening Up, the city launched a series of international festivals to facilitate cultural exchange. With sufficient accumulation of economic resources by the 1990s, Shanghai gradually restored its cultural infrastructure by constructing and renovating cultural venues that met international standards. Since the late 2000s, Shanghai's cultural sector has developed the systematic capacity to facilitate national and international cultural exchange and trade, especially of the performing arts, thanks to the increasing number of booking agencies, production teams, and other supporting services in the city (Zhang 2002).

Entertainment and fashion constitute most cultural consumption choices in Shanghai, synchronising the trends in cultural industries around the world (Jiang 2015). Shanghai is leading the national arts consumption trend in China. Arts and cultural training and activities, such as dance, painting, music, and theatre have witnessed a growing market in Shanghai. The 2010s saw intensive cultural exchanges within the region, and the discourse of *Jiangnan culture* (the cultural diaspora in South Yangtze River region) was recognised in Shanghai's most recent city branding agendas.

Since 2017, to fulfill the city's agenda of becoming the performing arts capital in Asia, there witnessed a new round of planning of the performing arts districts in the city center and the suburban districts. In addition to creating a liberalisation of cultural space through infrastructure development, the city also aimed to project a welcoming gesture "towards artists and creative talents from other cities and countries by promoting artistic collaboration and competition" (Kong 2015, p. 51). The "Shanghai Development Pattern" (p. 51) highlights the efforts of bridging artists and audiences through arts and culture-related experiential activities – the soft cultural infrastructure of the city.

City officials and planners have been granted more autonomy since the 2010 Shanghai EXPO, to creatively utilise local cultural assets and resources in city branding, corresponding to the practices of post-industrial cities in the west (Vanolo 2008). Carrying forward the inclusive agendas of community self-governance, social fabric regeneration, and human capital development, creativity has been widely adopted to enhance city images. Symbolic meanings and experiences are derived not only from material urban elements such as events and infrastructures, but immaterial ones such as rituals, institutions, and policies.

Shanghai's overall cultural atmosphere has been gradually led by its cultural and creative industries with the emphasis on services to cultural productions, and investments in cultural facilities, public culture, arts education and training, and event planning (O'Connor 2009; Lu 2018). It resonates with Charles Landry's (2005) ecological and comprehensive conceptualisation of the creative city through hardware and software infrastructure development. Through the supportive cultural policy environment, both "a culture of creativity" (p. 2) has been distilled to influence how urban stakeholders operate and an innovative and creative atmosphere is enabled for all urban residents.

CSIAF organisational structure – history, stakeholders, and programs

The organisational structure, mission, and aims of CSIAF

The China Shanghai International Arts Festival is hosted by the Ministry of Culture (now the Ministry of Culture and Tourism) as the only official state-level international arts festival in China. While not funding the festival, the Chinese government supervises the overall organisation and planning of the CSIAF. Although the festival relies on sponsorships and partnerships for funding opportunities, it is partially subsidised and organised by the Shanghai Municipal People's Government. After the first CSIAF in 1999, the Center for China Shanghai International Arts Festival (further referred to as the Center) was established to alleviate government bureaucrats' burden of managing the festival. Consisted of professionals from the field of event planning and arts management, the Center was created as a public cultural service unit with limited autonomy influencing festival programming and taking the full responsibilities of program implementation, together with partners and collaborators from the cultural sector in Shanghai and around the world.

Structurally, the Center is now overseen by the Shanghai Municipal Administration of Culture and Tourism and the Publicity Department of the Chinese Community Party (CCP), the Shanghai Municipal Committee. Like other public institutions in China, a Deputy Secretary of the General Party Branch of the Center oversees the public relation and arts education of the festival. Hence its programming and management are subject to politics and bureaucracies. Unlike the common board-based non-profit organisational structure in the west, the CSIAF does not have a board but an executive director, who is also the Branch Secretary of the CCP Party, coordinating general festival management and decision-making. The fact that the CEO serves as the Party Branch Secretary implies the heavy political intervention over the operation and administration of the Center. In effect, the CEO becomes the spokesperson not only of the festival itself but also of the CCP.

The artistic committee, the executive director of the Center, and the festival organising committee composed of politicians from state departments and city bureaus collectively decide the final performing arts programs. The implementation of the CSIAF is comparatively politics-free in comparison with the artistic program selection process, as festival programs are delivered through partnerships. The Center's staff team is quite small. Programming around 200 events and activities every year, the CSIAF operates on merely around 50 staff, supported by college interns and volunteers.

Although the Chinese version of the festival mission remains the same over the years, indicating it is by nature a celebration of arts and a festival for people. The English translation has been through changes. It has been settled down since 2015 on the succinct version "A Festival of Arts, a Gala for All", guided by the principles of "innovation and development" with an emphasis on brand-effect creation. It implies the festival as a local creative strategy for urban cultural regeneration, through introducing high-quality artistic performances to the local cultural venues and audiences and for city branding.

International events have been strategically used to integrate the domestic cultivation of citizenry and the international showcase of a better image of the host nation like the 2008 Beijing Olympics and the 2010 Shanghai EXPO. The creation of CSIAF coincided with the period when the Chinese government started to claim its global status through winning the bids of a variety of international events, such as the aforementioned Beijing Olympics (2000) and Shanghai EXPO (2002). According to Fang (2016), the CSIAF was founded with the explicit mission of facilitating international cultural exchange and establishing a platform to present local, national, and international arts and cultural works (Fang 2016). The festival hence bore similar agendas as the Shanghai EXPO, showcasing an emerging global power in response to the nation-branding agenda.

This annual festival provides a whole month of programs during October and November, with the original aim of introducing residents to the highest quality performing arts. Over its development, the CSIAF diversified its programs to not only open local audience's horizons but also make the city attractive internationally (Fang 2016). Co-presenting international and Chinese performing arts as the core programs, such as music, dance, theatre, and Chinese opera, the festival also features art forms like visual arts exhibitions. Additionally, affiliated programs like public cultural events and arts education activities associated with performing arts have been experimented and designed. Different local ethnic cultural forms and special themes are featured every year.

The history of CSIAF

Launching such a state-designated international arts festival in Shanghai in 1999 was inevitable. As China's financial capital, Shanghai had historically been the port for the cultural trade between China and the world, thus a perfect candidate for an international festival that is representative of contemporary China. The city has always been pioneering initiatives since China's Reform and Opening-up in 1978.

At the turn of the Millennium, the advanced development of Shanghai's physical infrastructure for performing arts also granted the city advantages to accommodate quality performing arts events. The increasing capacity and demands of residents in Shanghai to consume performing arts products created a market for an international arts festival. In the end, the comparatively free political and market environment has allowed the experimentation of a professionally managed arts festival in Shanghai.

According to Fang (2016), the increasing need for cultural exchange as an approach to opening China to the world and the fast development of China's economy contributed to the creation of the CSIAF. In its most recent mission statement, the festival continued to strive to become "a significant platform for cultural exchange and one of the leading arts festivals in the world".[4] Given the city managers' ambitions of revitalising Shanghai's global city status since Reform and Opening Up, the birth of the CSIAF was thus a collective effort between China's Ministry of Culture and Gong Xueping, the Deputy Secretary of Shanghai Committee of Communist Party of China. Programmed with the influences of both national and local agendas, the CSIAF leverages different aspects of Chinese culture and a variety of festival program types, under two CEOs[5] in its 21 years' history. Such programs as cultural exchange and spectacles (Hayden 2012) are adopted to cultivate and demonstrate the soft power of Shanghai and China domestically and globally (Nye 1990, 2005; Wuthnow 2008).

CSIAF stakeholders

Festival stakeholders take roles as producers, co-producers, facilitators, regulators, and the impacted participants and audiences (Getz 1997). In this study, the key CSIAF stakeholders are defined as individuals and groups that are publicising information related to the festival both directly and indirectly (see Table 3.1). These stakeholders can be categorised into the festival organisation – the Center, and its collaborators in the production of festival programs. Besides, the CSIAF's partners that facilitate festival implementation through presenting, marketing, funding, and sponsorship also play a role in promoting the festival. Their intentions are either associated with the festival brand or to obtain financial benefits. Hence the messages delivered by participating artists, presenting venues, and partnering institutions were examined, aside from sponsors' websites, government agencies' websites, and the festival's website (Fang 2016).

CSIAF Programs

In its first 13 years, the CSIAF developed a six-section program portfolio under the leadership of Chen Shenglai, the first executive director of the Center, who prioritised introducing quality international performing arts to Shanghainese. In addition to the core programming on performing arts showcases, the creation of the Trade Fair section since the first year of the CSIAF was a result of the Chinese government's expectation of the festival operated with the entrepreneurial and

Table 3.1 Categorisation of key CSIAF stakeholders

Stakeholder roles	CSIAF stakeholders	Stakeholders identified in this study
Organisers	The Center for CSIAF	The Center for CSIAF
Collaborators Co-producers	Local performing artists and groups	Shanghai Symphony Orchestra, Shanghai Philharmonic Orchestra, Shanghai Opera House, Shanghai Rainbow Chamber Singers, Shanghai Ballet, Shanghai Puppet Theatre, Shanghai Dramatic Arts Center, Shanghai Yue Opera Troupe, Shanghai Theatre Academy, Shanghai Art Theatre, Shanghai Changning District Centre for Shanghai Opera, Shanghai Huju Opera Theatre
	National performing artists and groups	Yang Liping, Ye Xiaogang, Tan Dun, Lei Jia, Jiang Yuequan, China National Opera House, National Theatre of China, The Palace Museum, National Ballet of China, Macao Orchestra, Inner Mongolia National Art Theatre National Music Orchestra, Hainan Centre for the Performing Arts, Ningbo Performing Arts Group, Edward Lam Dance Theatre, Sichuan People's Art Theatre, Shaoxing Xiaobaihua Yue Opera Troupe, Jincheng Shangdangbangzi Theatre, Ningxia Performing Arts Group, Jiangsu Performing Arts Group, Nanjing Repertory Theatre, Suzhou Su Opera Inheritance, and Preservation Center, Wuxi Opera & Dance Drama Theatre
	International performing artists and groups	Lv Jia, Riccardo Chailly, Ballet Nacional de Cuba, Anne-Sophie Mutter, Krzysztof Penderecki, Paavo Jarvi, Renaud Capucon, Alan Gilbert, Emmanuel Krivine, Roman Kim, Lucerne Festival Orchestra, Teatro Carlo Felice, Sinfonia Varsovia Poland, Juilliard String Quartet, Spanish Orfeo Catala, Tonhalle Orchestra Zurich, Camerata Salzburg, NDR Elbphilharmonie Orchestra, Tokyo Opera Singers, Orchestra National de France, Lithuanian Chamber Orchestra, Goteborgs Symfoniker Piano Quartet, Maria Pages Company, New York City Ballet, Les Ballets Jazz de Montreal, Polish Dance Theatre, The Australian Ballet, Sankai Juku, The Norwegian

(Continued)

Table 3.1 (*Continued*)

Stakeholder roles		CSIAF stakeholders	Stakeholders identified in this study
Organisers		The Center for CSIAF	The Center for CSIAF
		Presenting venues	National Ballet, Goteborgs Operans Danskompani, Akram Khan Company, Teatrul National "Radu Stanca" Sibiu, Kote Marjanishvili State Academic Drama Theatre, Romeo & Julia Koren, Dusseldorfer Schauspielhaus, Theatre des Bouffes du Nord Shanghai Grand Theatre, CADILLAC Shanghai Concert Hall, Oriental Arts Center, Majestic Theatre, Shanghai Hongqiao Arts Center, Yunfeng Theatre, Shanghai International Dance Center, SAIC-Shanghai Culture Square, Shanghai Symphony Hall, Daguan Stage, East Bund – Minsheng Wharf, STA Theatre, Shanghai Daning Theatre
		Partnering institutions	Shanghai Theatre Academy, kindergartens, primary and high schools in Shanghai, colleges, and universities in Shanghai, community cultural centers in Shanghai's districts, the municipal people's governments of Wuxi, Hefei, and Ningbo, the provincial government of Inner Mongolia
	Facilitators	Government agencies	China's Ministry of Culture and Tourism Shanghai Municipal People's Government Shanghai Municipal Administration of Culture and Tourism
		Media	Shanghai Broadcast and TV Station, People's Newspaper Corporation, Shanghai Newspaper Corporation, Xinhua.net, Tencent.com, Thepaper.com
		Sponsors	Volvo Cars, Bank of Shanghai, BOSCH, Shanghai Pharma, InterContinental Shanghai Jing'an, Happy Captain, Huiyuan Juice
The impacted	Participants and attendees	Ticket buyers, the public, students from primary and high schools, college students, participating artists, and professionals	Residents from Shanghai's districts, residents from surrounding cities of Shanghai and other Chinese cities, artists, arts groups, and professionals from the local cultural sector and international cultural and festival fields

Source: Own elaboration.

self-sustaining mode (Fang 2016). Putting the Performing Arts section at the core, the festival Exhibition section, Public Cultural Activities section, Performing Arts Trade Fair section, Forum section, and Themed Festivals section had been gradually added into the festival's program portfolio in the early phase of the CSIAF development (Chen et al. 2013).

In the later phase, frequent experimentations on new programs were particularly visible when commissioned works and young artists' works became important parts of the CSIAF program portfolio. Raising Artists Work (R.A.W.) was introduced in 2012 to cultivate young Chinese artists through workshops and to offer opportunities to present their works in the CSIAF. The Public Culture section split into the independent Shanghai Citizens Arts Festival and the Art Space section, aiming to bring cultural experiences into different types of public spaces. The increasing number of arts education activities evolved into its section – Arts Plus. The Themed Festival has developed into fruition by extending the impacts of the CSIAF to the whole city and the surrounding regions, through Cultural Weeks, Parallel Sessions, and collaborations with other international arts festivals in the city.

The CSIAF celebrated its 20th anniversary in 2018. This year was also the 40th anniversary of the nation's Reform and Opening Up since 1978. A special program "20●40 Coming Home" was designed to feature reputable Chinese artists who have been frequently performing at and contributing to the CSIAF in the past 20 years, a generation of artists growing out of the new era of development in China. A series of performances, public events, and exhibitions were dedicated to praising China's economic, social, and cultural achievements in the past 40 years (Table 3.2.). From October 19 to November 22, 2018, Art Space offered a variety of international and ethnic cultural experiences to a wide range of audiences in public spaces and community cultural centers. Some educational programs and inter-urban cultural exchanges were arranged during the spring and summer of 2018, such as student training workshops in Hong Kong and the new parallel session at Inner Mongolia.

CSIAF as a branding platform for China and the city of Shanghai

Branding contemporary China

As a platform for cultural exchange and presenting high-quality artworks from home and abroad, the festival fulfills the national branding needs. The Chinese government's branding efforts are focused on framing the image of contemporary China to its peers in the world. The overarching branding agenda for China can be summarised as "peaceful development" and to be perceived as a responsible country that protects ethics and morals when dealing with international affairs (He 2006; Wu 2009; Wu and Chen 2013). It is preferred that China carries an "orderly, prosperous and legitimate" (Barr 2012, p. 81) image.

The nation's cultural image is created through comprehensively presenting the historical, contemporary, and future China (Sun 2010; Men and Zhou 2013).

Table 3.2 Program portfolios of the 2018 China Shanghai International Arts Festival

Programs	Featured events in 2018	Types of activities
Performance	New York City Ballet Rainbow Choir 20–40 Coming Home	Showcasing performances by international and Chinese artists and arts groups
Exhibitions	Shen Wei: Exploring The Unkown	Free and discounted visual arts exhibitions
Art Space	Special Event of 12 Hours Forest Concert	Bringing diverse types of art forms from inland China and abroad (more western classical/ethnical culture) to local audiences at different venues and public spaces
Raising Artists' Work (R.A.W.)	6 commissioned work R.A.W.Land!	Select and workshop high-quality original works by emerging artists; Commissions and promotion the works at Trade Fair; R.A.W. land! offers more opportunities to showcase and learn from each other
Arts Plus	Shanghai-Hong Kong Youth Art Exchange Program Student Arts Jury Family Arts Camp	Campus tours; family sessions; training, exchange, conference, and workshops; guided viewing of performances
Trade Fair	R.A.W commissioned work pitch section Special Pitch Session for Chinese work going out Festival Investment session	Promoting a variety of works (from R.A.W. and Australian works to children's theatre and artworks from other regions in China); workshops;
Forum	Silk road Arts Festival Network Annual Conference Shanghai Global Cities Culture Forum 2018 Arts and Culture in the 21st Century The Development of Chinese Traditional Opera since 1998 The Development of Contemporary Visual Art in China Festival workshop for audience engagement and outreach	Themed conferences and workshops designed for professionals in the field from around the globe
Themed Festival	Jiangsu Cultural Week Parallel Sessions in Wuxi, Ningbo, Hefei, and Inner Mongolia Shanghai Puppet Festival Shanghai Baoshan International Folk Arts Festival	Cultural week; parallel session; themed festivals

Source: Author's elaboration.

The culture-related branding themes of China are multi-dimensional.[6] To improve Chinese people's cultural confidence, the nation needs to be united on shared cultural identities, built on the traditional and contemporary Chinese cultural assets. Aligning with the ideal modern cultural image of China proposed by Men and Zhou (2013), the Chinese should be presented as a civilised people who are well educated in Chinese culture and international cultures. Traditional and contemporary Chinese culture are major resources for national image framing and comprehensive national development, with the ambition of achieving the Chinese Dream (Cho and Jeong, 2008; Wang 2011).

The nation's cultural identity and cultural image contribute to the creation of China's cultural soft power, whereas international cultural exchange and parallel presentation of Chinese and international works further the promotion of China's soft power. Consequently, developing and demonstrating China's cultural soft power has been prioritised in a series of China's foreign and domestic policies, such as the Belt Road initiative, sustainable development, creative talent cultivation, national cultural security, and cultural system reform.

(Re)branding Shanghai as a global city

The Shanghai 2035 Master Plan proposed to make "A Charming City of Happiness and Humanity" through cultural engagement, social cohesion, cultural heritage preservation, and urban cultural landscape display. Specifically targeting residents, public and arts education programs in the CSIAF contribute to Shanghai's cultural development and urban regeneration agendas.

Presenting artworks by arts groups in Shanghai and the art forms originated from the city, the CSIAF also adds to the branding of Shanghai through symbolic meanings embedded within the selection of artistic contents in festival programs. Since the 1990s, the major city branding goal has gradually evolved to pursuing a global city status through "rebuilding and recovering a lost identity" (Schilbach 2013, p. 223) of modern Shanghai. The Shanghai Culture Brand strategy[7] was created based on a series of traits of Shanghai cultural identities and personalities, such as the openness to other cultures, professionalism in the cultural sector, cultural innovations, commercialisation of culture, and quality-oriented cultural productions.

Since the 2000s, culture has been moving upwards on local agendas[8] through urban ambitions such as making Shanghai an "international cultural exchange center", a "global cultural metropolis", a "creative design city", a global fashion city, and "the performing arts capital in Asia".[9] The event city, exhibition city, and tourism city agendas have been proposed by local intellects. In non-official media channels, Shanghai was also branded as the city of lights and the future, as well as the most beautiful city and the city that never sleeps. These endeavors resonate with the practices used in western city branding, i.e. the use of creative industries, culture, and festivals in brand creation (see Vanolo 2008 and Cudny 2019, pp. 29–35).

CSIAF as a place product

CSIAF and the festival ecology in Shanghai

Cudny (2019) distinguished two elements in place event marketing. One of them encompasses the creation of events as part of a city's overall cultural and entertainment offer to inhabitants, tourists, and other city users. In Shanghai, a large number of festivals constitute such a cultural offer. The super-sized city allows diverse types of festivals and events to make Shanghai an eventful destination for not only its citizens but tourists and potential investors. As one of the oldest arts festival in the city, the CSIAF is a critical part of Shanghai's festival ecology.

With its unique offerings, the festival sustains its status by offering high-quality artistic experiences to local and regional audiences, as well as national and international professionals, with different types of activities. The CSIAF also directly targets emerging artists' needs through cultivating and showcasing young artists. The trade fair section has always been credited as the most professional and reputable performing arts trade fair in the nation, in Asia, and even one of the must-attend trade fairs for performing arts professionals in the world. Cooperating and partnering with the cultural sector for program delivery, the festival adds to the development and improvement of the overall cultural system in Shanghai.

Over the past decades, CSIAF has contributed to the cultural infrastructure and atmosphere development in Shanghai, by bringing carefully programmed artistic experiences to different venues and public spaces. With another round of cultural infrastructure development in the city from 2016, CSIAF continued to be the content provider for newly constructed venues in suburban districts. The year 2018 witnessed performances presented at Shanghai International Dance Center, Shanghai Hongqiao Arts Center, and the East Bund – Minsheng Wharf.[10] As such, CSIAF is a valuable place product to broaden public access to arts and culture and contribute to Shanghai's eventful city, global city, and performing arts capital agendas.

The artistic experiences and symbolic meanings offered by CSIAF 2018 programs

With 3,500,000 audiences[11] attending over 350 activities in 2018,[12] different sections of CSIAF programs added to city branding through cultivated experiences, particularly towards the domestic audience. The 11 exhibitions covered a wide range of artistic forms, with a focus on contemporary arts from Chinese and international artists. Similarly, the total number of 66 performances in theatres was programed inclusively, in terms of artistic genres, themes, and the origins of performers.

The 2018 CSIAF presented less than 50% international works of the total performances, suggesting an emphasis on showcasing China's artistic achievements. Among the 39 Chinese repertoires, over two-thirds were the contemporary artistic interpretations of traditional Chinese artworks, stories of famous historical figures, and those of average people in the past 40 years of China's Reform and Opening Up. Juxtaposed with Chinese contemporary artworks, many international arts groups brought world-famous productions that adopted the contemporary or ethnic interpretations of western classics. The festival attempted to showcase China's cultural image not only through reinterpreting its traditional Chinese art forms but using contemporary and classical western art forms to tell Chinese stories.

These performances were brought by 34 international artists and arts groups, as well as 43 Chinese ones. The origins of international arts groups covered different continents and regions around the world, from Europe and the United States to Japan and Australia. The geographical origins of companies and artists encompassed a broader scale – from Lithuanian Chamber Orchestra, Estonian artist Paavo Jarvi, and Ballet Nacional de Cuba, to Korean artist Roman Kim, Camerata Salzburg, and the Australian Ballet. European countries constituted the majority of the homes of these touring artists groups, due to the artistic quality of their productions and the preservation of classical western cultural heritage in Europe.

Additionally, an increasing number of countries that participate in China's Belt Road initiative brought their performances to CSIAF in 2018, as part of the exchange between Belt Road countries. The performing arts fair and forum attracted professionals from within China and around the world to facilitate trade and knowledge exchange. The Chinese state thus presented itself as part of the global art field and a leader in the regional art market.

In contrast to presenting a large variety of China's ethnic cultures from around the nation as in previous editions of CSIAF, the year 2018 witnessed an emphasis on the cultural presentations of Shanghai and cities from the Yangtze River Delta region. This was particularly seen in the cities chosen to feature in Cultural Weeks and those chosen for Parallel Sessions. Performances by Shanghai's local arts groups constituted over a third of the 39 Chinese works presented in the 2018 CSIAF, in addition to 8 groups from the Yangtze River Delta region. Hence this festival highlighted the regional *Jiangnan Culture* and Shanghai's *Haipai Culture*[13] to demonstrate a unique aspect of Chinese culture.

Branding China and Shanghai were intertwined through the programming of the 2018 CSIAF, where the symbolic meanings of certain events were leveraged. The inter-urban exchange with China's inland region – Inner Mongolia was to showcase Shanghai's cultural production capacities and achievement. Besides, it brought ethnic culture from inland China to residents in Shanghai and the surrounding region, to celebrate cultural diversities and shared connections among different regions in China.

The special event "20•40 Coming Home" invited Chinese artists with global reputations back to the festival to share their experiences with CSIAF and showcase Shanghai's fast-developing cultural sector to the general public. Both the performances and exhibitions in this special section featured works by artists from Shanghai to reflect on the past 40 years of the Reform and Opening Up in China, celebrating the significant changes of the nation from the perspective of urban cultural regeneration.

Shanghai's comprehensive urban cultural development was further facilitated by the 2018 CSIAF, beyond simple city branding. Half of the five commissioned performing artworks were co-produced with artists and arts groups from Shanghai, which suggested the city's agenda of using this platform to cultivate the local cultural sector and creative talents. As a week-long arts carnival, the R.A.W. Land! section offered extra opportunities to showcase and train local young artists through the collaboration with Shanghai Theatre Academy.

Art Space and Arts Plus sections have a local focus to make arts and cultural experiences more accessible to cultivate future cultural consumers and artists, through the creative usage of public spaces and interactive activities. Partnering with Shanghai Puppet Festival, Shanghai Comedy Festival, and Shanghai Baoshan International Folk Art Festival, the 2018 CSIAF brought local audiences diverse cultural experiences and attracted visitors and tourists to the city.

The international–domestic perspectives of CSIAF branding of places

The international presentation of place brands

In addition to the place product creation, Cudny (2019) distinguished image creation through media as another element of the place event marketing process. Synthesising the international presentation of CSIAF in the media reports found on Google and promotional videos on YouTube, different branding themes of Shanghai and China emerged. There were significantly more messages used to promote Shanghai as a city than the national image towards an international audience.

Branding the images of Shanghai

Shanghai was portrayed with an image that is associated with its rich modern history and promising future of urban cultural development. Specifically, the city was presented as open to different cultures from local and regional areas within the nation and around the world, through the media exposure of the diverse local, national, and international arts and culture in the 2018 CSIAF programs. Additionally, the city was inclusive of emerging and established artists and artworks. The 2018 CSIAF's critical functions in cultivating young audiences and young artists were promoted through the heavy publicity of the Arts Plus program. It resonated with the city's long-term plan and efforts in developing original works by emerging artists and improving the city's creative capacity to achieve global city status.

Both YouTube videos and news articles on Google directed the audience's attention to Shanghai's urban cultural atmosphere and the city's overall physical infrastructure than the conventional branding of the downtown area. This was presented in the media by highlighting hundreds of public events taking place in suburban districts and unconventional spaces in Shanghai. Such an approach was vividly promoted as "theatres without the wall"[14] in the Art Space program.

The global city brand of Shanghai was implied in revealing the festivals' high standards of programming that matched other reputable international arts festivals. The professionalism and efficiencies in festival management suggested a well-developed cultural sector and urban governance capacity in Shanghai comparable with other international event host cities. The city was portrayed as offering a thriving cultural infrastructure, a growing cultural market, and a well-cultivated audience-ship for the successful delivery of CSIAF programs. Shanghai was thus framed as a city fully equipped with cultural assets and resources ready to become "an excellent global city"[15] by 2035.

The city was also constantly positioned as "the source" of quality artistic production in China in the media exposure of the 2018 CSIAF. It hence contributed to the promotion of the "Shanghai Culture" brand, depicting Shanghai as a model Chinese metropolitan city with the capacity of representing and producing the highest standard artistic achievement of the nation. The city's important status in China was highlighted in the media presentation of the 2018 CSIAF towards international audiences. The city was frequently referred to as "a port" connecting China and the world by facilitating international cultural exchange through the festival. By revealing the festival's spillover effects on the Yangtze River Delta region, Shanghai was branded as the leader in regional cultural development.

Branding the images of China

A smaller amount of messages in the media reports found on Google and videos on YouTube portrayed the Chinese state as a non-threatening player in global cultural affairs. Resonating with the nation's recent focus on cultivating patriotism, these messages appreciated the value of traditional Chinese arts and culture in contemporary Chinese society. Additionally, the Chinese cultural image was presented in the 2018 CSIAF comprehensively and authentically by including ethnic cultures from inland Chinese regions.

Juxtaposing the Chinese reinterpretation of Western classics and Chinese traditional arts with contemporary works from other countries, the nation's brand was made clear to its external audiences: Chinese culture shared commonalities with the rest of the world and contemporary China is comparable culturally with its peers. Showcasing original works by Chinese artists, both world renowned and emerging, CSIAF was used to reinforce the nation's "created in China" campaign. The nation's cultural confidence was reflected in the increasing number of export of original Chinese artistic works highlighted in the 2018 CSIAF trade fair section.

Another major theme in these international media channels was a forward-looking attitude with the expectations of the nation to continue its cultural development through cultural innovation, active engagement with the rest of the world, and cultivating young artists. A close read of these messages revealed the political agendas embedded in the promotion of Chinese artworks. The nation's cultural achievements were frequently attributed to its 40 years of development since

Reform and Opening Up. Hence, the nation attempted to promote *the Chinese road* as the development and political model to the world.

Yet the nation was very open with its political past, through the explicit reference to the *Red Culture*[16] in China's civil war period and the reinterpretation of stories from the Cultural Revolution period in the 2018 CSIAF's programs. It suggested the nation's efforts and confidence in recognising the country's comprehensive cultural identities. China's contemporary political agenda, especially the Belt Road initiative, was also highly visible. As such, China was not only branded as a confident and compatible player in the global cultural field but a leader for regional cooperation and co-development through Belt Road.

Chinese soft power was thus comprehensively constructed through presenting the wholeness and the advancement of Chinese culture, as well as making political statements of China's leadership in the Belt Road economic region through strategic festival programming. The diplomatic gestures of cultural exchange in the core program and knowledge exchange in the festival Forum and Arts Plus section resonated with the overarching theme of China's soft power agenda that emphasises the nation's peaceful rise through cooperative international relations.

The domestic presentation of place brands

There was a larger amount of media reports found on Baidu and video clips on Tencent Video. One significant characteristic of these news reports and videos was that they constituted primarily mainstream media channels, both locally and nationally. Consequently, Shanghai and China were branded through CSIAF following more prominent domestic cultural agendas. Place branding through the 2018 CSIAF towards China's domestic audiences was also more comprehensive than that towards international spectators.

Branding the images of Shanghai

The city's personalities were branded concerning its most recent cultural brand strategy,[17] which highlighted the importance of *Haipai Culture*, *Jiangnan Culture*, and *Red Culture* as the essence of Shanghai's cultural image. The city was heavily presented as resourceful in local cultural assets and cultural infrastructure, and ready with cultural agendas and policies.

Speaking to the *Shanghai Four Brands* Strategies initiated in 2017, the city was frequently referred to as "the source" of contemporary Chinese cultural production. As "the port" for the export of Chinese arts and culture, Shanghai was branded as a leader in regional development and a future leader in the Asia-Pacific region. The city was given a fuller image of being not only a leader but a collaborator, a coordinator, and a model, through increasingly frequent inter-urban exchange and resource sharing in the Yangtze River Delta region and the rest of China. Shanghai was hence depicted as a city capable of competing with other global cities culturally in the world.

In the promotion of the 2018 CSIAF, Shanghai's urban agendas were branded in the media with specific titles, such as the global city, the international cultural metropolitan, and the performing arts capital in Asia. Making these urban agendas explicit to the domestic audience contributed to a multi-faceted image of Shanghai that was recognizable by Chinese people.

The domestically oriented promotional messages focused more on specific local cultural agendas. Policies on cultural market development and creative talent cultivation were referred to when marketing the festival's R.A.W. section and Trade Fair. This was responding to Shanghai's intensive efforts in cultivating its creative sector, with an emphasis on attracting creative talents from home and abroad and developing creative communities (Keane 2009).

The domestic branding of Shanghai was people-centric. Through the heavy promotion of the discounted tickets program, Arts Plus, and Art Space programs, the public engagement of CSIAF and the city were highly visible. Besides, the active arts audience-ship in Shanghai was presented to create an authentic and festive city image. Such presentations of the festival and the city in the background not only contributed to Shanghai's Global City agenda but also added to its tourism city agenda. The 2018 CSIAF hence presented a creative urban living atmosphere attractive to the creative class, diverse cultural groups, and cultural investment.

Branding the images of China

Similar to the international presentation of China's cultural image, different ethnic Chinese cultures were promoted towards the domestic audience to present the diverse and complex aspects of Chinese culture. China was not merely identified with several well-developed cities along the coast, but the increasing growth of different inland regions.

Another theme consistent with the branding of China to an international audience was the support of original works created by emerging Chinese artists. Although these works made contemporary Chinese culture accessible to domestic audiences by telling Chinese stories with contemporary and global perspectives, they obtained less media exposure than the Chinese reinterpretation of Western classics. For the cultivation of national cultural confidence, the shared commonalities between Chinese culture and other cultures in the world were heavily addressed in the domestic branding of the nation through the 2018 CSIAF.

As the year 2018 was the 40th anniversary of Reform and Opening Up, the 2018 CSIAF presented not only the artistic achievement of China in this period but the future possibilities of China's artistic development. Consequently, though traditional Chinese culture was part of the CSIAF programs promoted towards domestic audiences, it was usually associated with "contemporary interpretations". Directed towards the domestic audience, the promotion of CSIAF served as a mobilisation tool for national agendas. President Xi's speech on cultural production in China in 2014[18] had been referred to frequently. Positive attitudes and the Chinese spirit of diligence were incorporated and promoted in the 2018 CSIAF programs, with the final goal of achieving the Chinese Dream.

The domestic branding of China covered an expanding array of policies on arts and culture. These agendas on culture were promoted straightforwardly to Chinese people by relating CSIAF programs with cultural policies so that the underlying logics of co-developing cultural industries and cultural affairs were accessible. Hence, Chinese people have presented comprehensive national cultural and political strategies, with more emphasis on domestic development than external cultural relations. The Chinese state was presented as a humble player in international affairs in its branding towards the domestic audience through CSIAF. The festival was promoted as a platform for the exchange and communication between China and the rest of the world, with an emphasis on China learning from its peers' development strategies and professional skills in arts and culture.

A comparison between the international and domestic presentations of place brands

In summary, Shanghai and China were branded differently through the 2018 CSIAF to international and domestic audiences. While the city was heavily branded towards the international audience, the nation branded towards the domestic audience. Though images of Shanghai and China were depicted similarly both internationally and domestically, they were presented with different emphases (see Figure 3.1). The festival only branded partial images intended by

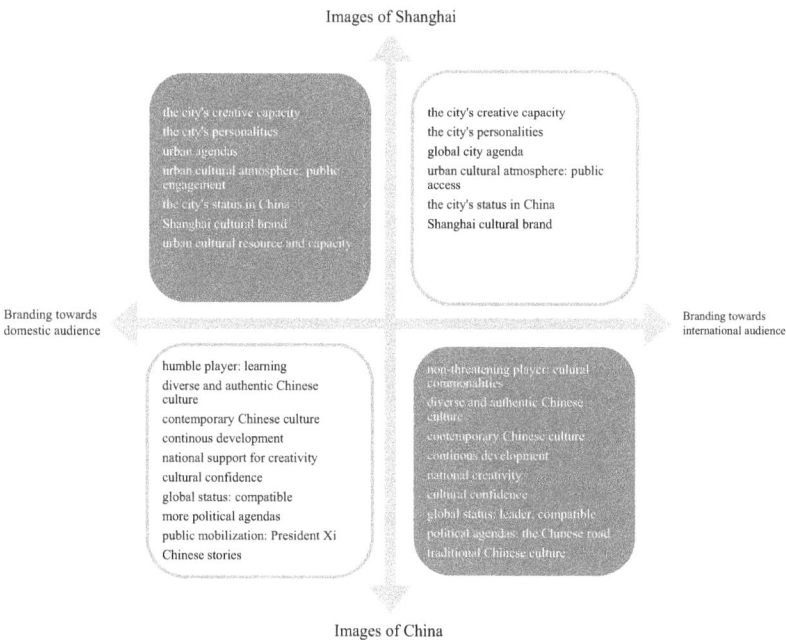

Images of Shanghai

the city's creative capacity
the city's personalities
urban agendas
urban cultural atmosphere: public engagement
the city's status in China
Shanghai cultural brand
urban cultural resource and capacity

the city's creative capacity
the city's personalities
global city agenda
urban cultural atmosphere: public access
the city's status in China
Shanghai cultural brand

Branding towards domestic audience

Branding towards international audience

humble player: learning
diverse and authentic Chinese culture
contemporary Chinese culture
continous development
national support for creativity
cultural confidence
global status: compatible
more political agendas
public mobilization: President Xi
Chinese stories

non-threatening player: cultural commonalities
diverse and authentic Chinese culture
contemporary Chinese culture
continous development
national creativity
cultural confidence
global status: leader, compatible
political agendas: the Chinese road
traditional Chinese culture

Images of China

Figure 3.1 Brand themes for Shanghai and China through place branding of the 2018 CSIAF towards international and domestic audiences. Source: Own elaboration.

the nation, in comparison with the branding themes of China identified in the previous section. There lacks the presentation of Chinese people's quality of life and the nation's responsibility in facilitating global governance. In contrast, Shanghai was given full attention in the branding of the 2018 CSIAF, responding to a full range of branding themes implied in the multiple urban cultural policy documents.

Co-constructing the place images through CSIAF and its stakeholders

Branding the places through the official CSIAF media

Among the 42 program-related news articles published on the festival's website,[19] a strong sense of branding Shanghai was expressed through publicising the festival's cooperation, exchange, and partnership with other Chinese cities or arts groups. The festival's coordination capacity suggested the city's leading role in China, especially in cultural production.

The English version of the official website of CSIAF aligned with the place branding patterns towards international audiences as discussed in previous sections. The city's cultural facilities, cultural resources, an increasingly professional cultural sector, a thriving cultural market, and a festival ecology were highlighted. Additionally, the "20•40 Coming Home" was promoted with a focus on the city's development in the festival's 20-years' history, attracting international attention to the overall cultural atmosphere in Shanghai.

The focus on promoting the host city of the 2018 CSIAF was intertwined with the constant reference to national cultural and economic development. The majority of works promoted in the news articles on the festival's website covered Chinese originals, commissioned works, and adaptations of Chinese traditional artworks. While the "20•40 Coming Home" special event was featured in the news, it is the development of the festival and Shanghai that was highlighted to demonstrate the nation's development in the past 40 years' Reform and Opening Up.

Branding the places through cultivated experiences at the CSIAF events

The experiential engagement offered by the 2018 CSIAF to participants helped to construct their perceptions of the city image, together with intense promotion and advertising of the festival through different communication channels. These cultivated experiences were not only created by the artistic experience through CSIAF programs but reinforced by the event atmosphere at the venues where festival programs took place. Different venues were used by the 2018 CSIAF to engage the audience, by offering performing arts experiences at conventional theatre spaces and unique experiences at unconventional public spaces, such as shopping malls, libraries, and parks. Shanghai was thus brought to the forefront by being presented as a city with an open and vibrant urban cultural atmosphere.

Scattered interviews with audiences and videos taken at Art Space programs[20] added another layer to the presentation of experiences offered by CSIAF. People's expectations of quality artistic experiences were satisfied. The videos taken from the site of the Gongqing Forest Park Music Festival[21] depicted an accessible and natural space. The open and interactive festival atmosphere and the respectful and appreciative audience-ship helped to construct an engaging festival-going experience. As the facilitator of such extensive, large-scale, and high-quality artistic experiences through CSIAF, Shanghai made a pleasant image of a city capable of coordinating its resources and personnel that meet the standards of other global cities.

Place images co-constructed by stakeholders and non-stakeholders of the 2018 CSIAF

Different stakeholders (and non-stakeholders) of the 2018 CSIAF presented the distinct festival values and place brands towards different audiences through a variety of messages in different media channels. This was most visible in the international branding of Shanghai and the domestic branding of both the city and the nation. Messages directed to international audiences were dominated by local and national mainstreams, the Center, tourism agencies, and artists associations (See Figure 3.2), whereas little branding efforts came from festival facilitators and co-producers. In comparison, aggregating the media messages from the festival's website and other media channels mentioned in previous discussions, the domestic presentations of the city and national brands were more comprehensive, covered by a diverse group of festival stakeholders and non-stakeholders.

The national and local mainstreams, along with the festival and its stakeholders composed complex and compensatory images of the nation and the city towards the domestic audience of the 2018 CSIAF. When branding Shanghai domestically, the national mainstreams delineated the city's various agendas, whereas local mainstreams focused on specific cultural images prioritised by Shanghai's most recent

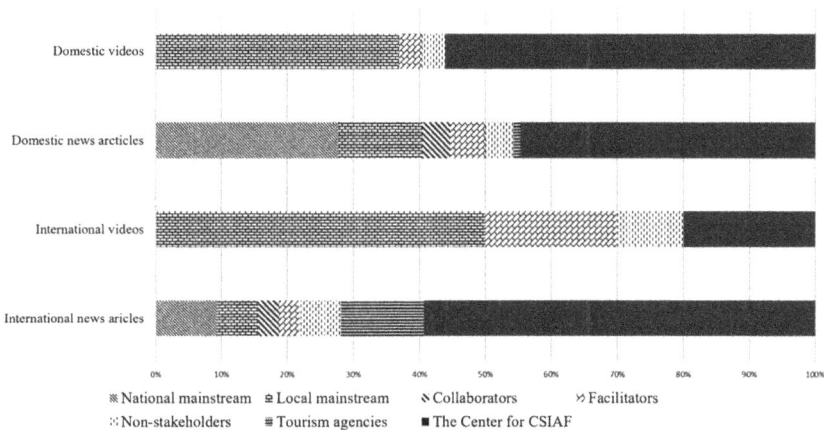

Figure 3.2 The percentage of messages by CSIAF stakeholders and non-stakeholders in domestic and international media channels. Source: Own elaboration.

urban agendas. Similarly, the national mainstreams highlighted China's overall national political agendas in the domestic branding of the nation. While local mainstreams promoted certain national strategies relatable to residents in Shanghai, such as the public cultural service system and the regional development agenda.

Therefore, CSIAF stakeholders (and non-stakeholders) addressed different aspects of the city and national brands. As cultural intermediaries, they have different communicative capacities and channels to frame the festival and the host city and nation to the general public, visitors, and other intended audiences. Hence, "their cooperation is a particular game of conflicting interests" (Cudny 2016, p. 88). Since reposting mainstream news articles had become a common practice for festival stakeholders and non-stakeholders to promote the 2018 CSIAF, the co-constructed images of Shanghai and China tended to be homogeneous than conflictual.

Conclusion

This chapter investigated the dynamics of city branding through festivals using a case from the Asia-Pacific region – the 2018 China Shanghai International Arts Festival (CSIAF). It addressed the different extents to which CSIAF was used by Shanghai to brand itself as a global city, and by the Chinese state to brand the nation's contemporary image through arts and culture. The following findings reveal the intricacies in the place branding practices through festivals.

CSIAF offered multiple venues of place branding. These venues include the creation of place product for the city of Shanghai, the symbolic presentations of national and local images in festival programs, and promotional messages via different media channels (including government websites, mainstream media, news reports, and commentaries of festival stakeholders and non-stakeholders, and the festival's official website). The experiential encounters at CSIAF reinforced some images preferred by the city and the nation and contributed to the overall place branding efforts. Resonating with Cudny's (2019) concept of place event marketing, CSIAF offered a valuable place product with multi-dimensional branding channels.

Given the unique political context in China, CSIAF served both nation branding and city branding agendas. Hosted by the Ministry of Culture and presented by the Shanghai Municipal People's Government, CSIAF was a tool to fulfill national and local agendas. The "image production" (Quinn 2005, p. 930) of cities is subject to different urban governance systems, hence the place branding responsibilities may distribute among different governments and non-governmental actors (Karvelyte 2017). The study suggested that the city branding agenda has been prioritised over national agendas in the 2018 CSIAF, especially in branding towards international audiences. This is due to the increasing autonomy of the city in planning and managing cultural affairs since the 2010 Shanghai EXPO and the rising urban global city agenda.

In the case of CSIAF, the place images were projected to both international and domestic audiences through place event marketing efforts. However, both Shanghai and China were branded with positive images preferred by the city and national governments (Cudny 2019). There was a drastic difference between the international and domestic promotions and presentations of the 2018 CSIAF and the host localities. Shanghai was heavily branded to international audiences through the festival to advance its global city agenda. While the nation was heavily branded to domestic audiences, sending an explicit message of a well-round national cultural and political agenda to realise the Chinese Dream.

The festival branding of places might not be consistent with the authorities' intended place brand images. The 2018 CSIAF incorporated multiple local and national agendas in its mission. However, given the limited time and capacity of the festival organisation, there were high and low priorities to be fulfilled through different types of programs. Adding to the constraints was the increasing number of partners the festival had in the year 2018, whose interests also influenced the presentation of CSIAF. As such, these stakeholders co-constructed the images of Shanghai and China for the festival's domestic and international audiences, together with the Center and local and national governments.

This research offers additional implications for the use of government-funded large-scale arts festivals as tools of place branding by Asia-Pacific cities and nations. First, place brand images need different actors to work collectively – so that there won't be conflicting messages conveyed through event products, media presentations, and other place branding projects. Secondly, successful place branding that meets authorities' intended agendas needs a long-term development plan involving compensatory place brand products and well-thought media communication strategies.

Notes

1 Sawe, Benjamin Elisha. (2018, November 2). The 10 Largest Cities in the World. Retrieved from https://www.worldatlas.com/articles/the-10-largest-cities-in-the-world.html, accessed on 09/05/2019.
2 See the 2019 census data at https://www.statista.com/statistics/466938/china-population-of-shanghai/, accessed on 10/01/2019.
3 See the 2010 census data at http://worldpopulationreview.com/world-cities/shanghai-population/, accessed on 10/01/2019.
4 Retrieved from http://www.artsbird.com/en/enaaf/enafjj/20120606/13530.html, accessed on 08/25/2019.
5 The first CEO was Chen Shenglai, who had experiences working in the media and news industries in Shanghai and was in the position from 2000 to 2011. Wang Jun took the position since 2012, with extensive experience in international cultural exchange and large-scale event planning, given her experience at the 2010 Shanghai EXPO.
6 Based on the thematic analysis of *the 13th Five-Year Plan for Social and Economic Development* and *the 13th Five-Year Plan for Cultural Development in China.*
7 The *Three-Year Action Plan for Shanghai Cultural Brand* by Shanghai Municipal Government was released in May 2018 http://www.shio.gov.cn/sh/xwb/n790/n792/n1038/n1051/u1ai17092.html, accessed on 08/10/2019.
8 Including Shanghai 2035 Master Plan, the 10th, 11th, 12th, and 13th Five-Year Plans for Shanghai's Economic and Social Development, and the city's 12th and 13th Five-Year Plans for Cultural and Creative Industries.
9 See *The Opinions on Accelerating Shanghai's Cultural and Creative Industries' Creative Development* released in 2017 at http://wgj.sh.gov.cn/node2/n2029/n2392/u1ai154359.html, accessed on 08/25/2019.

10 Shanghai International Dance Center was opened in October 2016, as the only cultural venue designed specifically for dance performances in Shanghai; Reconstructed from the former Tianshan Film Theatre, Shanghai Hongqiao Art Center become the new cultural landmark in Changning District in June 2016, with a 1,000-seats theatre and 7 cinemas; East Bund – Minsheng Wharf was transformed from the industrial heritage – Tubular Warehouses with the capacity of containing 80,000 tons by the wharf for previous Shanghai Port in 2017.

11 There are no available data regarding the composition of the tourists and local audiences attending CSIAF, but a speculation could be made. The festival drew mainly local audiences given its lack of international media exposure, according to an assessment report of the international media exposure of six major cultural events in Shanghai in 2016, accessed in Shanghai Library on June 24, 2019.

12 See the reports from Shanghai Municipal Government http://www.shanghai.gov.cn/nw2/nw2314/nw2315/nw5827/u21aw1351251.html, accessed on 09/20/2019.

13 See https://www.shine.cn/opinion/foreign-views/1806025684/ for an discussion of Haipai Culture, accessed on 10/20/2019.

14 More details about the Art Space program could be found at https://www.artsbird.com/NEWCMS/artsbird//cn/cn_17/cnystk_17/20151117/21601.html, accessed on 10/05/2019.

15 Shanghai Master Plan (2017–2035). Shanghai Scientific & Technical Publishers.

16 Red culture is an important and unique part of Chinese culture, referring to the early history of Chinese Communist Party (CCP) in the Chinese Revolution era (1921–1949), when the members of CCP, activists and the general public collectively created the Chinese socialist spiritual and ideological culture. It is regarded by CCP as the foundation of modernist Chinese civilisation, embracing the pragmatic development approach of contemporary China.

17 See *Shanghai Cultural Brand Three Year Action Plan (2018-2020)*, accessed on 10/05/2019, retrieved from http://www.shanghai.gov.cn/nw2/nw2314/nw3766/nw3895/nw43960/u1aw617.html.

18 See the whole speech at https://www.bbc.com/zhongwen/simp/china/2015/10/151015_china_xi_speech_literature_arts, accessed on 10/20/2019.

19 CSIAF website https://www.artsbird.com/ was accessed on 10/10/2019.

20 Interviews and videos retrieved from https://mp.weixin.qq.com/s/cChSbSBgoTgG2ffaoAuvFw, https://mp.weixin.qq.com/s/DtrVo0vBoIr9S7oHP5tBNQ and https://mp.weixin.qq.com/s/hUPNynpDGCIVATM8EWQfGQ, accessed on 10/28/2019.

21 Video retrieved from https://mp.weixin.qq.com/s/j-SLxTc-PFOPX89lfgAbYA, accessed on 10/28/2019.

References

Anholt, S. (2008). Place branding: Is it marketing, or isn't it?*Place Branding and Public Diplomacy 4*, 1–6.

Anholt, S. (2010). Definitions of place branding: Working towards a resolution. *Place Branding and Public Diplomacy*, 6(1), 1–10.

Arcodia, C., & Whitford, M. (2007, January). Festival attendance and the development of social capital. *Journal of Convention & Event Tourism*, 8(2), 1–18). Taylor & Francis Group.

Ashworth, G. (2009). The instruments of place branding: How is it done? *European Spatial Research and Policy*, 16(1), 9–22.

Ashworth, G., & Kavaratzis, M. (2009). Beyond the logo: Brand management for cities. *Journal of Brand Management*, 16(8), 520–531.

Ashworth, G.J., & Voogd, H. (1994). Marketing of tourism places: What are we doing?. *Journal of International Consumer Marketing*, 6(3–4), 5–19.

Barr, M. (2012). Nation branding as nation building: China's image campaign. *East Asia*, *29*(1), 81–94.

Bianchini, F., & Parkinson, M. (Eds.). (1994). *Cultural Policy and Urban Regeneration: The West European Experience*. UK: Manchester University Press.

Broudehoux, A.M. (2017). *Mega-events and Urban Image Construction*. Beijing and Rio de Janeiro. Routledge.

Chen, S., Ren, Y., Hua, J., Ruao, X., Xu, Q., Li, Y., … Shi, C. (2013). *Yishujie Yu Chengshi Wenhua* [*Arts Festival and Urban Culture*]. Shanghai: Shanghai Social Academy Publishing.

Cho, Y. N., & Jeong, J. H. (2008). China's soft power: Discussions, resources, and prospects. *Asian Survey*, *48*(3), 453–472.

Crang, M. (2003). Qualitative methods: Touchy, feely, look-see?. *Progress in Human Geography*, *27*(4), 494–504.

Cudny, W. (2013). Festival tourism–the concept, key functions and dysfunctions in the context of tourism geography studies. *Geographical Journal*, 65(2), 105–118.

Cudny, W. (2016). *Festivalisation of Urban Spaces: Factors, Processes and Effects*. Cham: Springer.

Cudny, W. (2019). The concept of place event marketing: Setting the agenda. In: Cudny, W. (Ed.) *The Concept of Place Event Marketing Setting the Agenda Urban Events, Place Branding and Promotion: Place Event Marketing* (pp. 1–24). London/New York: Routledge.

Dong, T. (2015). *Yishujie Yuanliu Yu Dangdai Fazhan Yanjiu* [*The History of Arts Festivals and Their Contemporary Development*]. (Thesis, Shanghai Theatre Academy).

Dwyer, C., & Beavers, K. (2011). *Economic Vitality: How the Arts and Culture Sector Catalyzes Economic Vitality*. Washington, DC: American Planning Association.

Falassi, A. (1987). Festival: Definition and morphology. *Time out of Time: Essays on the Festival* (pp. 1–10). Albuquerque, NM: University of New Mexico Press.

Fang, J. (2016). *Chengshi Jieri: Zoujin Zhongguo Shanghai Guoji Yishujie* [*City Festival: A Close Look at China Shanghai International Arts Festival*]. Shanghai: Shanghai Jiao Tong University Press.

Foley, M., McGillivray, D., & McPherson, G. (2012). *Event Policy: From Theory to Strategy*. London: Routledge.

Garcia, B. (2004). Cultural policy and urban regeneration in Western European cities: Lessons from experience, prospects for the future. *Local Economy*, *19*(4), 312–326.

Garcia, D.B. (2009). IMPACTS 08: European capital of culture research programme. Retrieved from https://www.liverpool.ac.uk/media/livacuk/impacts08/pdf/pdf/Creating_an_Impact_-_web.pdf

Getz, D. (1991). *Festivals, Special Events, and Tourism*. New York: Van Nostrand Reinhold.

Getz, D. (1997). *Event studies: Theory, Research and Policy for Planned Events*. New York: Cognizant Communication Corporation.

Getz, D. (2008). Event tourism: Definition, evolution, and research. *Tourism Management*, *29*(3), 403–428.

Getz, D. (2010). The nature and scope of festival studies. *International Journal of Event Management Research*, *5*(1), 1–47.

Gold, J.R. (2016). *Cities of Culture: Staging International Festivals and the Urban Agenda, 1851–2000*. London: Routledge.

Hanna, S., & Rowley, J. (2008). An analysis of terminology use in place branding. *Place Branding and Public Diplomacy*, *4*(1), 61–75.

Hayden, C. (2012). *The Rhetoric of Soft Power: Public Diplomacy in Global Contexts.* Washington, DC: Lexington Books.

He, H. (2006). Zhongguo guojia xingxiang dingwei fenxi [*An analysis of China's national image*]. *Contemporary Communication, 2,* 113–117.

Hildreth, J. (2010). Place branding: A view at arm's length. *Place Branding and Public Diplomacy, 6*(1), 27–35.

Jiang, L. (2015). Shanghai: Leading the National Art Consumption Trend. *Shanghai Economy, 06,* 28–30.

Kampschulte, A. (1999). "Image" as an instrument of urban management. *Geographica Helvetica, 54*(4), 229–241.

Karvelyte, K. (2017). Making a Creative City with Chinese Characteristics: Perspectives from Shanghai, Hong Kong and Taipei. (Doctoral dissertation, University of Leeds).

Kavaratzis, M. (2004). From city marketing to city branding: Toward a theoretical framework for developing city brands. *Place Branding, 1*(1), 58–73.

Kavaratzis, M., & Ashworth, G.J. (2005). City branding: An effective assertion of identity or a transitory marketing trick?. *Tijdschrift voor economische en sociale geografie, 96*(5), 506–514.

Keane, M. (2009). Creative industries in China: Four perspectives on social transformation. *International Journal of Cultural Policy, 15*(4), 431–443.

Kim, K., & Uysal, M. (2003). Perceived socio-economic impacts of festivals and events among organizers. *Journal of Hospitality & Leisure Marketing,* 10(3–4), 159–171.

Klaic, D. (2014). *Festivals in Focus.* Budapest: Regional Observatory on Financing Culture in East-Central Europe.

Klett, J. (2017). Reviews: Music/city: American festivals and placemaking in Austin, Nashville, and Newport, by Jonathan R. Wynn. *Contemporary Sociology, 46*(4), 496–497.

Kong, L. (2015). Rivalling Beijing and the world: Realizing Shanghai's ambitions through cultural infrastructure. In L. Kong, C. Chia-ho, & C. Tsu-Lung, *Arts, Culture and the Making of Global Cities* (pp. 49–63). UK: Edward Elgar Publishing.

Landry, C. (2005). *Lineages of the Creative City: Creativity and the City.* Rotterdam: Netherlands Architecture Institute.

Lee, I., & Arcodia, C. (2011). The role of regional food festivals for destination branding. *International Journal of Tourism Research, 13*(4), 355–367.

Lin, J. (2018). *Above Sea: Contemporary Art, Urban Culture, and the Fashioning of Global Shanghai.* UK: Manchester University Press.

Lu, X. (2018). Cultural industries in Shanghai: Policy and planning inside a global city. *Cultural Trends, 27*(5), 395–397.

Lucarelli, A. (2018). Place branding as urban policy: The (im) political place branding. *Cities, 80,* 12–21.

Mayring, P. (2004). Qualitative content analysis. *A Companion to Qualitative Research, 1,* 159–176.

Men, H. & Zhou, H. (2012). Zhongguo guojia xingxiang de jiangou jiqi chuanbotujing [*The creation of and communication of China's national image*]. *International Reivew, 1,* 8–15.

Nye, J.S. (1990). Soft power. *Foreign Policy, 80,* 153–171.

Nye, J.S. (2005). The rise of China's soft power. *Wall Street Journal.* Retrieved from https ://www.wsj.com/articles/SB113580867242333272

O'Connor, J. (2009). Shanghai moderne: Creative economy in a creative city? In L. Kong & J. O'Connor (Eds.), *Creative Economies, Creative Cities* (Vol. 98, pp. 175–193). Dordrecht: Springer.

Ooi, C.S., & Pedersen, J.S. (2010). City branding and film festivals: Re-evaluating stakeholder's relations. *Place Branding and Public Diplomacy*, *6*(4), 316–332.

Quinn, B. (2005). Arts festivals and the city. *Urban Studies*, *42*(5–6), 927–943.

Richards, G., & Palmer, R. (2010). *Eventful Cities: Cultural Management and Urban Revitalisation*. Amsterdam: Elsevier/Butterworth-Heinemann.

Riffe, D., Lacy, S. and Fico, F. (2014). Analyzing Media Messages: Using Quantitative Content Analysis in Research . New York/London: Routledge.

Ruzinskaite, J. (2015). *Place Branding: The Need for an Evaluative Framework* (Doctoral dissertation, University of Huddersfield).

Sacco, P., Ferilli, G., & Blessi, G.T. (2014). Understanding culture-led local development: A critique of alternative theoretical explanations. *Urban Studies*, *51*(13), 2806–2821.

Schilbach, T. (2013). Cultural policy in Shanghai: The politics of caution in the global city. In *Culture and the City* (pp. 34–48). London: Routledge.

Sun, Y. (2010). Zhongguo guojiaxingxiang de wenhuajiangou [*The cultural construction of China's national image*]. *Teaching and Research*, *11,* 33–39.

Thomas, J., & Harden, A. (2008). Methods for the thematic synthesis of qualitative research in systematic reviews. *BMC Medical Research Methodology*, *8*(1), 45.

Todd, L., Leask, A., & Ensor, J. (2017). Understanding primary stakeholders' multiple roles in hallmark event tourism management. *Tourism Management*, *59*, 494–509.

Vanolo, A. (2008). The image of the creative city: Some reflections on urban branding in Turin. *Cities*, *25*(6), 370–382.

Wang, J. (Ed.). (2011). *Soft Power in China: Public Diplomacy through Communication*. Berlin: Springer.

Ward, S. (2005). *The Garden City: Past, Present and Future*. London: Routledge.

Waterman, S. (1998). Carnivals for elites? The cultural politics of arts festivals. *Progress in Human Geography*, *22*(1), 54–74.

Wilson, J., Arshed, N., Shaw, E., & Pret, T. (2017). Expanding the domain of festival research: A review and research agenda: Domain of festival research. *International Journal of Management Reviews*, *19*(2), 195–213.

Wong, P.F. (2014). *Shanghai, China's Capital of Modernity: The Production of Space and Urban Experience of World Expo 2010* (Doctoral dissertation, University of Birmingham).

Wu, W. (2004). Cultural strategies in Shanghai: Regenerating cosmopolitanism in an era of globalization. *Progress in Planning*, *61*(3), 159–180.

Wu, Y. (2009). Duiwai Wenhua Chuanbo yu Zhongguo Guojia Xingxiang Suzhao [External cultural communication and China's national image creation]. *International Review, 01,* 8–15.

Wu, F. & Chen, Y. (2013). Zhongguo guojia xingxiang pingshu [An overview and discussion of China's national image]. *Contemporary Communication, 01,* 8–11.

Wuthnow, J. (2008). The concept of soft power in China's strategic discourse. *Issues & Studies*, *44*(2), 1–28.

Yin, R.K. (2017). *Case Study Research and Applications: Design and Methods*. New York: Sage.

Zenker, S., & Beckmann, S.C. (2013). Measuring brand image effects of flagship projects for place brands: The case of Hamburg. *Journal of Brand Management*, *20*(8), 642–655.

Zhang, T. (2002). Urban development and a socialist pro-growth coalition in Shanghai. *Urban Affairs Review*, *37*(4), 475–499.

4 Events across ASEAN

Product-oriented regeneration and value-added image promotion

Nicholas Wise

Introduction

Events help places attract attention and media publicity, giving destinations a chance to display product and service offerings to international audiences (Stevenson 2013; Wise and Harris 2019). In this regard, events are increasingly playing a central role when it comes to place marketing and promotion (Cudny 2020). According to Cudny (2020), place event marketing is a strategic approach to diversifying event portfolios, planning future events, and communicating opportunities to wider audiences so that the events industry grows and is sustained over the long term. In relation to this chapter, place event marketing is about place development and destination image, with a regional consideration of product-oriented regeneration that in turn helps us understand value-added image promotion. Building on this focus is the need to also consider, alongside place marketing, how events also spark geopolitical debates, surrounding power relations, ideas of inclusion and exclusion, scalar relations, and the role of everyday interactions in reproducing social, political, and economic processes (Waitt and Gibson 2009; Wise 2017). From another perspective, spatial land-use planning for new stadium venues also impacts local populations, through both positive and negative development efforts (Wise and Harris 2017), as well as selecting who will host an event (Wright 2018), each aiding additional insights into place marketing and event promotion.

Traditionally many countries would identify the largest or capital city based on influence, infrastructure, and ease of transport connections, but there is an increasing trend to hosting events in lesser-known cities or regions as a way of using events to brand a destination and/or regenerate place images (see Lee 2017, 2019; Wise and Harris 2019). While mega- and large-scale sporting events are becoming increasingly popular, they are considered as drivers of development and/or redevelopment planning and city branding strategies (Knott et al. 2015; Koch 2013; Wise 2020), they have been controversial recently in terms of the amount of spend and criticised because public monies could be directed elsewhere (Lauermann 2019; Wise and Whittam 2015). Other scholars argue that there is also a need to focus on the role of events across wider geographical regions—including a range of different types of events and new infrastructures, spaces and places that enhance the capacity to host events, as evident in a recent collection by Wise and Harris (2019). While bidding, planning, and delivering most one-off multi-sporting events is a venture of one host-city there are new attempts to extend the impact of large-scale events to include multiple hosts so that several cities can benefit and a regional brand can emerge with an enhanced service offering that is moving towards a collective and sustained agenda.

Large-scale events are mass gatherings attracting a significant international audience (Wise and Harris 2017), whereas smaller-scale events seek to engage communities and promote a new sense of community (Zhao and Wise 2019). Given the regular occurrence of events, Jackson (2013, p. 847) noted: "the expanding annual calendar of sport mega-events, both in terms of the bidding process and the actual hosting of the event, means there is rarely time for considered reflection". Mega and major events are described as large-scale mass cultural gatherings attracting a significant international audience (Horne and Manzenreiter 2006; Roche 2000; Rojek 2013; Wise 2017). While sporting mega-events such as the Summer Olympic Games and the Men's Fédération Internationale de Football Association (FIFA) World Cup attract much attention, other large-scale or hallmark events also require further consideration as these also play a significant role in international expansion and place development. It is such events used by destinations and places that create new opportunities to develop a destination, before, during and after a large-scale or hallmark event. Moreover, there is a particular focus on sporting mega-events when looking at wider impacts on cities and regions, and while this chapter will bring in examples of major sporting events, it mainly aims to consider the wider role of events and how they impact destination marketing and branding from both regeneration (product-oriented) and place image promotion (value-added) perspectives.

This chapter will lead with conceptual perspectives and understandings guided by the literature on regeneration, (place) development, place image and destination branding before addressing considerations of competitiveness to frame these different but overlapping conceptual components of product-oriented and value-added development. This chapter will frame these conceptual directions and contexts by looking at the role and influence of hosting events in the ASEAN (Association of Southeast Asian Nations) region. While it is not possible to cover all recent events and developments in this region, examples within each conceptual area will highlight important issues and considerations. Considering the links across theory and practice is useful for event managers and urban/regional policy makers to consider as they continually use events to brand destinations or enhance place images as well as achieve regeneration agendas (especially in the larger urban centers across the region). One of the major factors here is regional competitiveness, and the need to promote a regional vision for events so that the region can collectively gain from the influence of events and the bourgeoning service economies that they create. However, if the market is over-saturated, this can in turn damage these eventful images and reputations because it in turn means that destinations may undersell themselves amid increased regional competition.

ASEAN region

ASEAN is an "intergovernmental organisation aimed primarily at promoting economic growth and regional stability among its members" (World Economic Forum 2020, online) comprised of 10 member states: Brunei, Cambodia, Indonesia, Laos, Malaysia, Myanmar, the Philippines, Singapore, Thailand, and Vietnam. In 2017, ASEAN celebrated 50 years of regional commitment, with a mission of "one vision, one identity, one community" (Wood 2017). In 2015, the region looked to define their future vision and the role of hosting large-scale events continues to be a focus as the region looks to build its international influence outside the region to promote cultural values and encourage business development (ASEAN 2015).

Economically and politically, the ASEAN region has established and negotiated free trade agreements among the 10 countries, have eased travel and mobility for citizens of these countries and have built connections with China given their proximity to the world's fastest growing economy. According to the World Economy Forum (2020, online), "if ASEAN were a country, it would be the seventh-largest economy in the world, with a combined GDP of $2.6 trillion in 2014. By 2050 it's projected to rank as the fourth-largest economy". The same source adds: "The ASEAN region is home to more than 622 million people, the region has a larger population than the European Union or North America. It also has the third-largest labour force in the world, behind China and India". Another focus and purpose of ASEAN is to promote regional peace and stability. There is also a collective effort to promote research cooperation and technological development to preserve the environment so that conservation efforts will lead to sustainable futures (ASEAN Heritage Parks is a joint effort looking after 37 protected sites/areas). In terms of socially dedicated efforts, there is a focus on education (through the ASEAN University Network founded in 1995) to promote cooperation and scholarship among youth, part of this was the University Games, a bi-annual event that commenced in 1981.

Given this regional growth and vision to define their place globally, cities across the region are investing in venues and facilities to attract people to the region using events as a driver to brand and re-brand destinations, enhance place images and build local, regional and national tourism economies, capacities and competencies. While a number of cities have hosted sporting events that have promoted inter- and wider-regional travel in Asia (e.g. the Asian Games held in Jakarta and Palembang in 2018), one of the key areas that ASEAN cities are investing in are conferencing and exhibition spaces to build a large-scale business or Meetings, Incentives, Conferences and Exhibitions (MICE) events industry. This is widely regarded as the largest and most sustainable form of events to invest in because they are regularly held, as opposed to most large-scale sporting events that are often times one-off events. If each country and region across ASEAN develop a regionally focused event that celebrates local culture, and brand that as a yearly hallmark event, this is a chance to highlight the diversity of the wider region and see a regular flow of participants. This is also more cost-effective as annual events can build on past successes and expand accordingly, larger-scale events that happen once require a lot of upfront initial investment and can result in over-supply, and can be an issue if the demand cannot be maintained later. For instance, events such as Brunei Gastronomy Week, Water and Moon Festival and Boat Racing in Cambodia or the Harvest Festival in Myanmar put these countries in a position to use events as a celebration that embraced opportunities and aligns with local and regional traditions that can also be enjoyed by new attendees and can help build a new tourism economy, thus expanding regularly held local events.

The larger cities, and arguably the most influential in ASEAN, are in competition with each other to attract larger business and MICE events. Cities that stand out in the region, investing heavily in venues and complexes include Singapore, Kuala Lumpur, Jakarta, Bangkok, and Manilla. This is not the first level of competition that these cities face, they are also the base for a number of global airlines, and these cities also compete for air traffic passengers travelling globally, so investments in modern airports with a wide range of transit amenities put these cities in direct completion. Singapore (Singapore Airlines), Kuala Lumpur (Malaysian Airlines), Jakarta (Garuda Indonesia), Bangkok (Thai Airways), and Manilla (Philippine Airlines) are in direct competition for passenger traffic, and other national carriers including Vietnam Airlines and Royal Brunei Airlines are gaining market presence, with Vietnam Airlines expanding rapidly in international markets. Myanmar, Laos, and Cambodia are arguably more peripheral in such considerations of development in event venues and large national airline carriers, but cities in these countries are well connected to hubs in fellow ASEAN countries. Despite the heightened competition among ASEAN countries to attract events and air passengers, these cities and national air carriers also need to compete with major cities in China (such as Guangzhou, hub of China Southern Airlines), as well as Hong Kong (hub of Cathay Pacific and Cathay Dragon), and Macau (a major entertainment center). While this chapter suggests that a region-wide strategy is necessary to develop events for the purpose of holistically marketing more remote regions and smaller cities, the current COVID-19 pandemic will disrupt these plans and recommendations (at least in the short term). In the immediate future, the larger cities with air transit hubs and large event venues are strategic for restating an event-oriented economy.

A number of examples will be highlighted in the following sections to relate the development of events in ASEAN alongside conceptual discussions of regeneration, place image, and competitiveness to address product-oriented and value-added perspectives of destination marketing and branding in Asia. Examples from across the ASEAN region support and offer some practical insight to complement the conceptual discussions.

The product-oriented focus: events, regeneration and (place) development

Regeneration and place development is a widely debated holistic concept among social science, development, and management scholars (e.g. Spirou 2011; Smith

2016; Wise, 2018). The topic has been the focus of research across a range of disciplines. Some studies looked at peoples' outlooks, attitudes, impact, and support, whilst other works focus more on economic conditions and encouraging cohesive involvement in communities. The development literature is concerned with spatial planning and transformations (e.g. Jones 2002; Thornley 2002), whereas entrepreneurial and management perspectives focus on encouraging people to create new enterprises (Hall 2006; Richards et al. 2013). A clearer focus that looks at the role of development in internationalising destinations, and framing this alongside national endeavors and sporting geopolitics that enable a place to host an event, in an attempt to develop events in a city and region is what is needed from a policy standpoint. Given the global endeavors to host sporting-events there is a need to focus on change and the pressure on places to keep up with the shifts in demand in the global economy so to maintain a competitive advantage. A point that is still accurate today is Richards and Palmer's (2010) argument that places the need to keep up with the pace of global change or they risk stagnation and decline as increased competition is resulting in new products and more opportunities and choice for consumers.

Development is the process of change (also referred to as regeneration, renewal, or revitalisation among scholars in different countries). One of the main motivations of regeneration is to create new opportunities in destinations for investors, stakeholders, and local communities. Around the world, hosting sporting events is driving development in cities and regions. Investments in sport, events, and subsequent tourism opportunities play a role in image transition and transitioning economic bases (Richards and Palmer 2010; Cowan 2016). Planning events as part of wider urban, regional, and national development strategies is a way of defining a place and its strategic importance (see Poynter et al. 2016; Tallon 2013; Wise 2016). Place development has become an important and contested concept because spatial transitions can have an impact on the local population as well as external perceptions (Spirou 2011; Wise and Whittam 2015).

These external pressures can refer to marketing dilemmas, as places seek to promote a destination or a new sports product in a highly saturated and competitive market. While the focus on sport involves global competition, across Asia we are seeing different regions also compete to host large-scale events or build large conferencing venues to help publicise destinations (see Lee 2017; Buathong and Lai 2017). Thus, places undergo extensive physical regeneration projects as a way of keeping up with the status quo, as city/regional planners, private investors, and practitioners deem them necessary to support economic development, competitiveness, and the chance to host future events and increase tourism (Wise 2019). But regional competitiveness is a major factor today, and there can be for instance underlying tensions and/or geopolitical pressures behind hosting and promoting international events so that one destination can maintain a competitive advantage.

New stadia, arenas, or conferencing venues not only represent (physical) infrastructural change but they contribute to new city images (Smith 2005; Vanolo 2015). Much work has addressed how image contributes to new city brands or new ways of welcoming tourists. A point relevant to Indonesia (and cities across

Asia hosting large-scale events) is Gover's (2011, p. 227) argument that destination branding "should be about creating an overarching brand strategy or competitive identity that reflects a particular nation's, city's or region's history". Over the past 20 years, cities have transformed their urban imagesas a way to overcome any past negative images or associations linked to economic decline, deindustrialisation, and/or regional restructuring (Wise and Harris 2017).

Countries across the ASEAN region of Asia greatly vary in terms of geography, population, and economic influence. Despite these noted geographical differences and factors, each country is attempting to transform cities by investing in events (at all scales and types, including sport and major exhibitions) to lever future benefits that will enhance the image of destinations, play a key role in modernisation and grow the service/tourism economy, and thus enhance marketing and branding strategies going forward. The challenge here is opposed to having one place that highlights sport and events, there is increasing competition among regions as many cities are adopting similar strategies. An over-saturation here, and a nation-wide shift towards sport and event service-based economies, is resulting in what some might argue over-festivalisation, and urban development that can threaten future growth in cities across the ASEAN region as they are continually competing to attract tourists and event-goers.

The focus of urban development and destination branding is often on income generation and reviving or further sustaining the local and urban economy (Spirou 2011). Economic indicators drive change and development, and increasingly today events are one of these indicators that can help to publicise places. Economic development in its broadest sense not only considers urban income generation, but also how such developments create new cultural, social, and employment opportunities for residents (Soh and Yuen 2011). Both public and private sector developments are done on the basis of long-term goals, sustainable planning, and gaining returns on stakeholder investments. However, outcomes assessed and future consequences are dependent on present-day decisions—therefore changing trends are often based on speculation. Going back to the focus on sporting events and development, we are seeing a shift to the East and South as cities in North and West are not seeing the long-term benefits of one-off sporting events (Maharaj 2015; Wise and Hall 2017). Thus, relationships between sporting mega-events, host cities, and associated development strategies have developed a burgeoning cross-disciplinary literature with discussions that are almost entirely shaped by potential legacies of mega-events, including increased sports participation, social benefits through transformation, and (possible) economic returns on investments by host nations and cities (Wise and Hall 2017; Wise 2019).

Much work has looked primarily at Western developed nations but a shift in hosting mega-events is turning attention to developing emerging economy countries (see Hall and Wise, 2019). While Western developed countries may no longer see the sustained benefits, sporting-events are seen as legitimate strategies and catalysts for social and economic development in emerging economies such as India, China, South Africa, and Brazil (Darnell 2012; Wise 2020). But when the focus shifts to social development which is widely criticised

(Coakley and Souza 2013; Hall and Wise 2019), this is something that urban and regional policy makers across the ASEAN region need to consider if they are to promote socially sustainable futures alongside larger-scale events. Moreover, such events serve to enhance these nation's power, economic competitiveness, and prestige in global relations (Maharaj 2015) with the presumed outcome of attracting international investment, positive media, and increased tourism (Curi et al. 2011). Broadly, however, there are still no satisfactory conclusions for a clear positive development and legacies that occur for nations or cities hosting sporting mega-events (Coakley and Souza 2013), despite considerations that need to be addressed in Asia highlighted over a decade ago (see Close et al. 2007; Dolles and Söderman 2008). Given the strong desire for cities across the ASEAN region to construct venues and host events (Henderson 2015), it is important that planners and those who market cities compare their situation and potential to sustain with that of other cities who have also (recently) hosted events.

There is a growing focus on case in different emerging economy countries; for instance, Wise and Hall (2017) focus on Brazil's attempt to develop their international events profile. In this case, until 2007, Brazil had not hosted any mega or major international sporting events since the 1950 FIFA World Cup and the 1963 Pan American Games. The 44-year hiatus ended following a series of successful bids to host the 2007 Pan American Games and the 2014 FIFA World Cup. The 2016 Summer Olympics in Rio de Janeiro closed out Brazil's sporting decade (see Gaffney 2010; Reis et al. 2013). Event-led and still on-going development initiatives in Brazil focus on developing cities and regions away from the coast as part of the country's attempt to modernising peripheral cities and regions (see Gaffney 2008; Wise 2019). To reiterate the point made about economic and social development outside a nation's dominant cities, Brazil nominated host cities in more peripheral regions. It can be a challenge to host larger-scale events in peripheral regions due to transportation and supplementary services that are necessary and a consideration that needs to be addressed is places may suffer later if the short-term demand has no long-term impact.

Across the ASEAN region, many second-tier and even smaller cities are looking to host larger events (e.g. Palembang as co-host of the 2018 Asian Games; Chon Buri, Thailand, will host the 2025 Southeast Asian Games). But smaller (or lesser-known internationally) cities do not always benefit from continual hosting, and this presents an issue because the larger (capital) cities in these countries (Singapore, Kuala Lumpur, Bangkok, Manilla, Jakarta, Ho Chi Minh City) have a strategic advantage being major global airline hubs with well-developed (often much better developed), inter-urban, and immediate regional transportation networks. There remains a concern around widespread event developments in smaller markets as this can result in a challenge for cities, and can over-saturate market demand for events. This in turn can threaten the growth and sustainability of events' economy in the longer term, as more venues means more choice and more competition to bid or attract events and this can mean that they need to undersell themselves. Major urban centers do not need to worry as much because they have economic (and political) influence as well as larger (and oftentimes much more modern) facilities.

The value-added focus: events, place image, and destination branding

Research on place and destination branding is well established in tourism marketing (e.g. Baker 2012; Kladou et al. 2016) and destination development literature (e.g. Dinnie 2015; Zenker and Jacobsen 2015). Such work is also seeing an increased focus in the urban and regional studies literature (see Cudny 2019; Hultman et al. 2016), especially because places seek to have a lasting impact on public perception (Salman 2008). This chapter is concerned with bringing these approaches together because branding a destination contributes to place marketing and promotion as a form of value-added for a place (Kozak and Baloglu 2011), and builds on the product-oriented considerations of place marketing and development addressed in the previous section of this chapter. Destination branding and destination image are two conceptually different areas of research, despite there being some clear links and associations (see Qu et al. 2011). The notion of branding differs from image, because image is a situational condition that is important to assess when considering how a destination is perceived – and is important when considering communication, consumer demand, and ability to attract visitors. From a psychology standpoint, brands and images are associations, framing associations with places, particular events, or points in history (Bassols 2016; Wise and Mulec 2015).

Place branding is important to consider because it highlights a particular awareness aligned with a particular vision of geographical areas (see Tiwari and Bose 2013). Aligned with the above discussion on regeneration, Cudny (2020, p. 1) argues: "branding in the case of cities is most often understood as a multidimensional urban development strategy" and further adds "the development of a city brand should lead to the creation of an attractive offer, which is often called a city product". Events regeneration has been referred to as events-led, events-themed, or events-focused, but given the increasing importance of events for cities to attract consumers, cities are seeking to brand destinations around events, which in turn can also enhance place image (Smith and Fox 2007).

Arguably this is much more straightforward when we consider a larger-scale event such as the Asian Games. Jakarta and Palembang as hosts of the 2018 Asian Games benefitted from the exposure, including in some of Asia's larger travel markets such as China, Japan, and Korea. Not only do larger-scale events benefit from media broadcasts, they gain from the spectacle created and the lasting effects of the event as a brand image. Nevertheless, beyond larger-scale events, smaller or even annual hallmark events are also gaining presence as places are branding more traditional events in a manner that increases their visibility regionally (or internationally). Some examples of such events from ASEAN include the Elephant Festival in the Sayaboury Province of Laos each February, Deepavali celebrations among Hindu and Indian communities across Malaysia and Singapore, Brunei's

Gastronomy Week, or Thailand's VERY Festival that gains the likes of sponsors such as Singha Corporation.

Destination branding aligns with destination marketing and promotion, to inform the place/destination offer (Kozak and Baloglu 2011). Similarly, corporations use branding to attract customers using recognizable logos, as we see with Singha and the VERY Festival in Thailand. Destination branding therefore, according to Kladou et al. (2016), refers to naming or developing recognizable logos or taglines to identify a destination, which can play a role in image regeneration. Likewise, Hanoi Pride, which saw its first parade in 2012, is Vietnam's first pride event and an opportunity for the LGBT community in Vietnam to use the event as a way to change social attitudes and policies nationally and regionally (see Oosterhoff et al. 2014). Hanoi Pride resembles similar events to help give the event more publicity to promote the event and the awareness it intends to create. The Legacy Festival in Singapore's Sentosa Island blends event, place, and destination branding. The slogan "A New Dawn" positions the nighttime experience of the event; it works as place promotion for Sentosa Island and promotes Singapore's destination/city brand as the "Lion City" all as part of the marketing package on websites and social media advertisements.

Destination images, alternatively, comprise of numerous components, including popular attractions, accessibility, facilities, and infrastructures. Baloglu and McCleary (1999) note that place and destination images involve personal factors (values, age, and particular motivations/interests) and stimulus factors (information sources, previous experiences), whereas branding is promoting "consumption" factors (product, appeal, marketing). There is a clear difference here in the types of events promoted. For example, sporting events seem to focus more on youth and experience (in the case of Jakarta and Palembang using cartoon images to appeal to and inspire younger attendees), whereas business events are strict and professional (in the case of the Philippine International Convention Center which promotes a more formal atmosphere or events and galas on its website). Factors offer working knowledge of who travels to particular destinations, and disseminated experiences encourage people to go out and create their own experiences (Camprubí et al. 2013), and part of the appeal is promoted through branding. Deepavali in Malaysia and Singapore received much attention with communities linked to the Indian diaspora, and this has likewise emerged into significant local celebrations in both countries.

Competitiveness: over competition or a regional vision?

Wise and Armenski (2020, p. 1) argue that "the contribution of events to destination revitalisation and competitiveness is an area of research that needs more attention, especially as events can have either lasting or contesting legacies". Wise and Harris (2017, 2019) add that place image and destination branding have been issues of increasing international importance for decades now, and this continued focus here on competitiveness frames the importance of driving new visitor economies using events against back-drops of economic uncertainty, demographic change, and technological innovation (see also, Smith 2016). In the 1990s, the Asian Tiger economies came to an abrupt halt with the 1997 Asian financial crisis that started in Thailand with the collapse of the Thai Baht currency; this caused a shockwave in the region and countries in ASEAN were significantly impacted very negatively by the sudden economic recession (Kaufman et al. 1999). In terms of competitiveness, Singapore has an advantage. The small city-state has a high quality of life and a high-income economy. Singapore stands out from the perspective of investing in events space. The city planned and built the south of the Marina Bay area for events, completed in 2010. By building and branding a "new" downtown area for hosting events, this gave Singapore a unique and strategic advantage to show their presence as an international host (Gwee 2013). While Singapore designated a particular space for event, Indonesia, which sits in the shadow as a major-events host, used the Asian Games co-hosted by Jakarta and Palembang, seen as a chance for Indonesia to promote the growth and increasing significance of the country in ASEAN and Asia (Davies 2014).

Destination competitiveness is a well-researched area, especially in tourism (e.g. Crouch 2011; Crouch and Ritchie 1999; Dwyer and Kim 2003), with more work recently looking at the impact of events (see Aquilino et al. 2019; Wise 2020). Linked to the context of this chapter, there needs to be not only a place-focused consideration to destination competitiveness but also a regional focus because no research has considered the development of a regional events portfolio because destinations are continually competing with each other as opposed to having a regional vision or strategy. ASEAN is a regional brand that connects members through this shared identity, but each city and the examples discussed here are operating independently to attract major events and conferences. Considering the examples of different events across ASEAN, each destination provides a wide-product offering, but there is a need for each country to perhaps focus on a key hallmark event that stands out across the region and internationally. Myanmar, for instance, does not host many events that promote an international audience, but one approach is to embrace the local approach to event planning. Leveraging local culture and promoting local authentic experiences is a competitive advantage. The Golden Hilltop Festival for instance has a long tradition of connecting locals from surrounding villages who converge upon the Kyaik Khauk Pogoda to buy and sell local products, and this event is a chance to experience local life, traditions, and celebrations just outside Yangon.

While each country may/will host similar events, if ASEAN planners can create a regional events calendar where different annual events are promoted, then this could be a step towards developing a regional strategy for promoting events. Moreover, this would align with the mission of ASEAN from a creative and cultural sector standpoint, and could also help reduce competition so that a regular events cycle emerges and could help create event tourism flows through the region. Then ASEAN countries promote the region as a whole opposed to an over-consumption approach based on competition. Competition will continue of course for events such as conferences and exhibitions, but the focus on achieving a holistic strategy whereby each country promotes a particular hallmark event from each country is a way to capture and achieve regional integration, regional celebrations, and regional spectacle. Part of the spectacle refers back to place images and power relations, whereby event venues as symbols of power convey to international audiences the increasing importance of events, this is seen in the case of Beijing (Broudehoux 2010) but is an increasing trend across Asia and within ASEAN as the cities have an advantage.

What we are seeing around ASEAN is many second-tier and smaller cities also building significant complexes and hosting annual hallmark events to build event-hosting capacity outside the larger/capital cities. Take for instance the Big Mountain Music Festival in Thailand; this event that takes place in Khao Yai proximate to a national park is the largest outdoor festival in Southeast Asia, attracting crowds of more than 25,000. Because the festival is held several hours from Bangkok, this spreads exposure and promotes hosting in more places outside the core (capital) cities. Now the countries that dominate the events industry use their product-oriented successes to add value to other nations who need extra assistance when it comes to promoting events and building a visitor economy. This is a form of events capacity building and reinforcing links whereby best practices are also disseminated which is another form of added value from a future marketing, promotion, and branding standpoint.

Here, the notion of spectacle aligns with image, which remains a popular topic of inquiry regarding perception, awareness, and knowledge of attractions, places, and nations. Considering music festivals in Indonesia, this has created event spectacles aimed at attracting younger travelers. As these festivals have gained in international exposure and popularity in recent years, the hosting of the Asian Games in two cities was a chance to use a major-event to leverage opportunities for the wider events industry. We The Fest has been considered a "pioneer" festival because it overlaps art, music, fashion, and food. While this appears a hodge-podge, this diverse approach allows festival-goers to have blended experiences and is viewed as an accommodating approach to planning events. We The Fest is held in Jakarta, but in line with the point above about spreading event hosting geographically, the LaLaLa International Forest Festival in Bandung and the Air Festival in Gili Air are extending popular events beyond Jakarta and using music as a strategy to enhance the marketing of music festivals across Indonesia.

An important conceptual link joining this focus on regeneration, image change, and competitiveness is scale, and some ASEAN nations are emerging as core players economically and politically, such as Indonesia (G20 member nation). Only little research to date has addressed events and scalar relations looking at where power and governance are concentrated, based on established hierarchies (Wise 2017). In ASEAN, Indonesia is emerging rapidly, but it is really the core cities that have invested in platforms and venues to build a sustainable events profile, especially Singapore's Marina Bay South area developed to host event or the expansive >100,000 sq. ft. Philippine International Convention Center in Manila. It must be noted that there exist similar sized convention centers across ASEAN, for instance in: Jakarta, Tangerang, Ho Chi Minh City, Bangkok, Pattaya, Bali, Nonthaburi, Kuala Lumpur, Kuching, Ayer Keroh, Kota Kinabalu, and Pasay.

Concerning events and scale in what is relevant from a regional context is that there are two divisions, which appear as scalar spectrums of cores and peripheries, and the semi-periphery, aligned with Wallerstein's (1974) world-systems theory. One way to consider this in ASEAN is through cities that hold higher-profile events and the countries that are not as competitive. For instance, Laos, Myanmar, and Cambodia as peripheral from a regional standpoint, Vietnam, the Philippines, and Brunei Darussalam are arguably semi-peripheral while the other countries have strategic advantages in terms of hosting events. This said, Singapore, Malaysia, Indonesia, and Thailand see their (capital) cities continually seeking opportunities to upgrade facilities, promote modern amenities, and world-class hospitality, this giving them an advantage in the bidding process. Singapore, Malaysia, and Thailand also have a strategic advantage given air transit access and connectivity with more connections, but Vietnam is looking to increase their global connections.

Some nations in ASEAN have or are rapidly emerging, growing or gaining new influence. Within this context, a focus on scale and scalar relations is a consideration that is often discussed by economic and political geographers who assess globalisation—and this links to the international expansion of hosting mega-events. According to Smith (2000, p. 724), scale refers to "one or more levels of representation, experience and organisation of geographical events and processes". Geographical and regional scales therefore align to competitiveness because there always exist political and economic complexities of power, control, and hierarchies (Wise 2017) which play a role in what nations and cities can invest and place attention on enhancing image, opportunities, and capacities.

With the regional commitment of ASEAN, what the "core" event hosting nations can do is help build closer ties so that influence and opportunities can be more dispersed. However, this might be difficult to realise given the rise of Singapore and Malaysia and more recently Indonesia, who are looking to invest future in larger venues. This has positioned these destinations as key global competitors as event hosts. Competing at a global scale can mean that these are the premier ASEAN host cities that can put regional members in a position where they are not competitive, and thus need to compete with the other smaller economies to attract smaller events, but these might lack the investments that destinations such as Singapore can achieve. Scalar shifts are dependent upon economic diversification, and while the region has been able to attract manufacturing and industry and to contribute to the global supply chain, the countries that invest in events are aiming to grow the service section as a way to attract further basic economic activity through visitor spend. This increased spend means these destinations can invest in attractions and bids, which leads to the establishment of a place brand, new spectacles and increased cultural influence and this is where large-scale events are significant.

Concluding remarks

The points offered in the above sections align with Dwyer and Kim's (2003, p. 369) point that to achieve a competitive advantage that meets appeal, a place "must be superior to that of the alternative destinations open to potential visitors". But the point being made here is to try and limit this alternativeness regionally by offering distinctly different hallmark events to grow the region's wider competitiveness. What can then happen is such a distinct regional events appeal can create/re-create a new image (for the region and for each place where the event is held, and can help promote less travelled destinations). There is also a need to forego any negative associations linked to developments in ASEAN nations, as sometimes these happen rapidly but miss promotion windows. Thus, and hosting a range of large-scale events, of culturally focused hallmark events can attract media attention to help transform the image of these countries, cities, and attractions—which can help shape future tourism legacies as well (Wise and Mulec 2015). However, it must be noted that large-scale events are not so regular, but the promotion of smaller MICE-oriented events can create a wider impact in terms of marketing destinations.

Looking at the future and direction of events in ASEAN, the most sustainable path is more specifically MICE events given the abundance of infrastructures in place (as cities with large conferencing and exhibition centers noted above). While these venues are scattered around the region, the five cities that stand out as leaders in event delivery and promotion in this chapter are Singapore, Kuala Lumpur, Jakarta, Bangkok, and Manilla. The advantage these cities have is they are core capital cities of their respective countries, and so they have the power to sustain place marketing strategies through investment in "national" venues and being hub cities for major airlines they have a benefited from this strategic advantage and the ability to attract visitors and host events with large crowd capacity.

Going back to the points about working towards a regional strategy to promote events, the COVID-19 crisis may put a delay in these plans. However, this region-wide approach is not instantaneous, and needs to be considered as a necessary long-term strategy (looking to develop events over the next several decades). If national governments of the ASEAN region want to extend the impact and promotional capabilities of events to grow this industry in cities and regions, then a plan to build appropriate sized venues and support bids to host more small- and medium-sized events is required. This is one way forward for lesser-known areas to build a new service-based economy and subsequently promote tourism, then these places can further invest in marketing capability region-wide beyond the key cities that gain from more exposure. If events in this region can support the growth of tourism and place product development, then MICE events are key because they attract not only larger numbers of attendees, but business events and conferences are opportunities for short breaks and many attendees travel with family members who in turn contribute to the tourism economy.

Once concern that Crouch (2011) addresses is the difficulty to manage rapid change. The issue is rapid change can increase competitiveness because places then need to rapidly respond to changes and need to respond with an alternative or similar offer. This is especially true going forward, when it comes to marketing and planning for events post-COVID-19 because of uncertainty over the desire to gather in large crowds and air travel. Going forward this will be an issue for smaller cities, especially. The five cities noted in the previous paragraph from a post-pandemic standpoint have large venues so they will be able to market events that will meet future spacing and social distancing requirements. Moreover, and being capital cities, each with strategic airline hubs, any reduction in flight demand will still see these cities operating as they handle flight connections through each city.

From a regional product development standpoint, it is ideal to look at the events offering, identify a particular hallmark event for each country, and begin to promote then as a regional events calendar so that places can build on the success of the event and regional events over time and work towards a strategic product and promotion strategy. Doing this can help countries have a clear focus on end goals because there is a desire to meet consumer appeal and demand, and an events calendar can help increase inter-regional mobility and international travel from outside the region. ASEAN nations each have differing motives, which may involve vastly different financial (of social and environmental) demands and consequences. A strategy forward is to work towards a clearly defined hallmark event in each country held on regular intervals through the year so that an events rotation sees each country providing a unique offer without compromise and direct competition that instead displays inclusion among all nations and a diverse offering of events.

References

Aquilino, L., Armenski, T., & Wise, N. (2019). Assessing the competitiveness of Matera and the Basilicata Region (Italy) ahead of the 2019 European Capital of Culture. *Tourism and Hospitality Research*, 19(4), 503–517.

Association of Southeast Asian Nations (ASEAN) (2015). ASEAN 2025: Forging ahead together. Available at: https://asean.org/?static_post=asean-2025-forging-ahead-toge ther (accessed 8 February 2020).

Baker, B. (2012). *Destination Branding for Small Cities*. Portland, OR: Creative Leap Books.

Baloglu, S., & McCleary, K.W. (1999). A model of destination image formation. *Annals of Tourism Research*, 26(4), 868–897.

Bassols, N. (2016). Branding and promoting a country amidst a long-term conflict: The case of Colombia. *Journal of Destination Marketing & Management*, 5(4), 314–324.

Broudehoux, A-M. (2010). Images of power: Architectures of the integrated spectacle at the Beijing Olympics. *Journal of Architectural Education*, 63(2), 52–62.

Buathong, K., & Lai, P-C. (2017). Perceived attributes of event sustainability in the MICE industry in Thailand: A viewpoint from governmental, academic, venue and practitioner. *Sustainability*, 9(7), 1151.

Camprubí, R., Guia, J., & Comas, J. (2013). The new role of tourists in destination image formation. *Current Issues in Tourism*, 16(2), 203–209.

Close, P., Askew, D., & Xu, X. (2007). *The Beijing Olympiad: The Political Economy of a Sporting Mega-Event*, London: Routledge.

Coakley, J., & Souza, D.L. (2013). Sport mega-events: Can legacies and development be equitable and sustainable? *Motriz: Revista de Educação Física*, 19, 580–589.

Cowan, A. (2016). *A Nice Place to Visit: Tourism and Urban Revitalization in the Postwar Rustbelt*. Philadelphia: Temple University Press.

Crouch, G.I. (2011). Destination competitiveness: An analysis of determinant attributes. *Journal of Travel Research*, 50, 27–45.

Crouch, G.I., & Ritchie, J.R.B. (1999). Tourism, competitiveness, and societal prosperity. *Journal of Business Research*, 44, 137–152.

Cudny, W. (2019). *City Branding and Promotion: The Strategic Approach*. London: Routledge.

Cudny, W. (ed) (2020). *Urban Events, Place Branding and Promotion: Place Event Marketing*. London: Routledge.

Curi, M., Knijnik, J., & Mascarenhas, G. (2011). The Pan American Games in Rio de Janeiro 2007: Consequences of a sport mega-event on a BRIC country. *International Review for the Sociology of Sport*, 46, 140–156.

Darnell, S.C. (2012). Olympism in action, Olympic hosting and the politics of 'sport for development and peace': Investigating the development discourses of Rio 2016. *Sport in Society*, 15, 869–887.

Davies, W. (2014). Indonesia to host 2018 Asian games. *Wall Street Journal*, 27 September, Available at: https://www.wsj.com/articles/indonesia-to-host-2018-asian-games-1 411189198 (accessed 10 February 2020).

Dinnie, K. (2015). *Nation Branding: Concepts, Issues, Practice*. London: Routledge.

Dolles, H., & Söderman, S. (2008). Mega-sporting events in Asia: Impacts on society, business and management: An introduction. *Asian Business & Management*, 7, 147–162.

Dwyer, L., & Kim, C. (2003). Destination competitiveness: Determinants and indicators. *Current Issues in Tourism*, 6, 369–413.

Gaffney, C. (2008). *Temples of the Earthbound Gods: Stadiums in the Cultural Landscapes of Rio de Janeiro and Buenos Aires*. Austin: University of Texas Press.

Gaffney, C. (2010). Mega-events and socio-spatial dynamics in Rio de Janeiro, 1919–2016. *Journal of Latin American Geography*, 9, 7–29.

Govers, R. (2011). From place marketing to place branding and back. *Place Branding and Public Diplomacy*, 7(4), 227–231.

Gwee, J. (2013). *Case Studies in Public Governance: Building Institutions in Singapore*. London: Routledge.

Hall, C.M. (2006). Urban entrepreneurship, corporate interests and sports mega-events: The thin policies of competitiveness within the hard outcomes of neoliberalism. *Sociological Review*, 54, 59–70.

Hall, G., & Wise, N. (2019). Introduction: Sport and social transformation in Brazil. *Bulletin of Latin American Research*, 38(3), 265–266.

Henderson, J. (2015). The new dynamics of tourism in South East Asia: Economic development, political change and destination competitiveness. *Tourism Recreation Research*, 40(3), 379–390.

Horne, J., & Manzenreiter, W. (eds). (2006). *Sports Mega-Events: Social Scientific Analyses of a Global Phenomenon*. Malden, MA: Blackwell Publishing.

Hultman, M., Yeboah-Banin, A.A., & Formaniuk, L. (2016). Demand- and supply-side perspectives of city branding: A qualitative investigation. *Journal of Business Research*, 69(11), 5153–5157.

Jackson, S. (2013). Rugby World Cup 2011: SportmMega-events between the global and the local. *Sport in Society*, 16(7), 847–852.

Jones, C. (2002). The stadium and economic development: Cardiff and the Millennium Stadium. *European Planning Studies*, 10, 819–829.

Kaufman, G.G., Krueger, T.H., & Hunter, W.C. (1999). *The Asian Financial Crisis: Origins, Implications and Solutions*. Berlin: Springer.

Kladou, S., Kavaratzis, M., Rigopoulou, I., & Salonika, E. (2016). The role of brand elements in destination branding. *Journal of Destination Marketing & Management*, 6(4), 426–435.

Knott, B., Fyall, A., & Jones, I. (2015). The nation branding opportunities provided by a sport mega-event: South Africa and the 2010 FIFA World Cup. *Journal of Destination Marketing & Management*, 4(1), 46–56.

Koch, N. (2013). Sport and soft authoritarian nation-building. *Political Geography*, 32, 42–51.

Kozak, M., & Baloglu, S. (2011). *Managing and Marketing Tourist Destinations*. London: Routledge.

Lauermann, J. (2019). The urban politics of mega-events: Grand promises meet local resistance. *Environment and Society*, 10(1), 48–62.

Lee, J.W. (2017). Mega-event skepticism in South Korea: Lessons from the 2014 Incheon Asian Games. In N. Wise and J. Harris (eds). *Sport, Events, Tourism and Regeneration* (pp. 40–53). London: Routledge.

Lee, J.W. (2019). A winter sport mega-event and its aftermath: A critical review of post-Olympic PyeongChang. *Local Economy*, 34(7), 745–752.

Maharaj, B. (2015). The turn of the south? Social and economic impacts of mega-events in India, Brazil and South Africa. *Local Economy*, 30(8), 983–999.

Oosterhoff, P., Hoang, T-A., & Quach, T.T. (2014). *Negotiating Public and Legal Spaces: The Emergence of an LGBT Movement in Vietnam*. Institute of Development Studies, 74, 1–44.

Poynter, G., Viehoff, V. & Li, Y. (eds). (2016). *The London Olympics and Urban Development: The Mega-Event City*. London: Routledge.

Qu, H., Kim, L.H., & Im, H.H. (2011). A model of destination branding: Integrating the concepts of the branding and destination image. *Tourism Management*, 32(3), 465–476.

Reis, A.C., Sousa-Mast, F.R., & Vieira, M.C. (2013). Public policies and sports in marginalised communities: The case of Cidade de Deus, Rio de Janeiro, Brazil. *World Leisure Journal*, 55, 229–251.

Richards, G., & Palmer, R. (2010). *Eventful Cities: Cultural Management and Urban Revitalisation*. London: Elsevier.

Richards, G., de Brito, M., & Wilks, L. (eds). (2013). *Exploring the Social Impacts of Events*. London: Routledge.

Roche, M. (2000). *Mega-Events and Modernity*. London: Routledge.

Rojek, C. (2013). *Event Power*. London: SAGE.

Salman, S. (2008). Brand of gold. *Guardian*, 1 October 2008. Available at: https://www.theguardian.com/society/2008/oct/01/city.urban.branding (accessed 8 February 2020).

Smith, A. (2005). Conceptualizing image change: The reimagining of Barcelona. *Tourism Geographies*, 7, 398–423.

Smith, A. (2016). *Events in the City*. London: Routledge.

Smith, A., & Fox, T. (2007). From 'event-led' to 'event-themed' regeneration: The 2002 Commonwealth Games legacy programme. *Urban Studies*, 44(5–6), 1125–1143.

Smith, N. (2000). Scale. In R. Johnston, D. Gregory, G. Pratt and M. Watts (eds). *The Dictionary of Human Geography* (pp. 724–727). Oxford: Blackwell.

Soh, E.Y.X., & Yuen, B. (2011). Singapore's changing spaces. *Cities*, 28(1), 3–10.

Spirou, C. (2011). *Urban Tourism and Urban Change: Cities in a Global Economy*. London: Routledge.

Stevenson, N. (2013). The complexities of tourism and regeneration: The case of the 2012 Olympic Games. *Tourism Planning & Development*, 10, 1–16.

Tallon, A. (2013). *Urban Regeneration in the UK*. London: Routledge.

Thornley, A. (2002). Urban regeneration and sports stadia. *European Planning Studies*, 10, 813–818.

Tiwari, A.K., & Bose, S. (2013). Place branding: A review of literature. *Asia Pacific Journal of Research in Business Management*, 4(3), 15–24.

Vanolo, A. (2015). The image of the city, eight years later: Turin, urban branding and the economic crisis taboo. *Cities*, 46, 1–7.

Waitt, G., & Gibson, C. (2009). Creative small cities: Rethinking the creative economy in place. *Urban Studies*, 46, 1223–1246.

Wallerstein, I. (1974). *The Modern World System*, Vol. 1. New York: Academic Press.

Wise, N. (2016). Outlining triple bottom line contexts in urban tourism regeneration. *Cities*, 53, 30–34.

Wise, N. (2017). Rugby World Cup: New directions or more of the same? *Sport in Society*, 20(3), 341–354.

Wise, N. (2018). Tourism and social regeneration. *Social Sciences*, 7(12), 262.

Wise, N. (2019). Towards a more enabling representation: Framing an emergent conceptual approach to measure social conditions following mega-event transformation in Manaus, Brazil. *Bulletin of Latin American Research*, 38(3), 300–316.

Wise, N. (2020). Eventful futures and triple bottom line impacts: BRICS, image regeneration and competitiveness. *Journal of Place Management and Development*, 13(1), 89–100.

Wise, N., & Armenski, T. (2020). The contribution of events to destination revitalisation and competitiveness. *Journal of Place Management and Development*, 13(1), 1–3.

Wise, N., & Hall, G. (2017). Transforming Brazil: Sporting mega-events, tourism, geography and the need for sustainable regeneration in host cities. In N. Wise and J. Harris (eds). *Sport, Events, Tourism and Regeneration* (pp. 24–39). London: Routledge.

Wise, N., & Harris, J. (eds). (2017). *Sport, Events, Tourism and Regeneration*. London: Routledge.

Wise, N., & Harris, J. (eds). (2019). *Events, Places and Societies*. London: Routledge.

Wise, N., & Mulec, I. (2015). Aesthetic awareness and spectacle: Communicated images of Novi Sad, the Exit Festival and the event venue Petrovaradin Fortress. *Tourism Review International*, 19(4), 193–205.

Wise, N., & Whittam, G. (2015). Editorial: Regeneration, enterprise, sport and tourism. *Local Economy*, 30(8), 867–870.

Wood, J. (2017). ASEAN at 50: What does the future hold for the region? World Economic Forum, 9 May 2017. Available at: https://www.weforum.org/agenda/2017/05/asean-at -50-what-does-the-future-hold-for-the-region (accessed 8 February 2020).

World Economic Forum (2020). What is ASEAN? Available at: https://www.weforum.org /agenda/2017/05/what-is-asean-explainer/ (accessed 8 February 2020).

Wright, R.K. (2018). Event bidding, politics, persuasion and resistance. *Annals of Leisure Research*, 21(5), 637–640.

Zenker, S., & Jacobsen, B.P. (2015). *Inter-Regional Place Branding: Best practices, Challenges and Solutions*. Berlin: Springer.

Zhao, Y., & Wise, N. (2019). Evaluating the intersection between "green events" and sense of community at Liverpool's Lark Lane Farmers Market. *Journal of Community Psychology*, 47(5), 1118–1130.

5 Brand association with a participant sporting event

The case of the Okinawa Marathon in Japan

Yosuke Tsuji and Carolin Schlueter

Introduction

Over the past two decades, the number of running events in Japan, especially road races such as 5-km runs, 10-km runs, and marathons, has increased. It is estimated that close to 2,800 running events are held each year in Japan (Takai, 2018). The phenomenon is driven mainly by the trend toward being health-conscious among the Japanese. Inoue (2018) argues that the start of the Tokyo Marathon in 2007 inspired people to take up running as their hobby. According to the Sasakawa Sports Foundation research, in Japan, the number of runners who run at least once a year has reached 10 million in 2012 (Inoue, 2018). However, the same report estimates that the number peaked in 2012 and has been slowly declining since then. Takai (2018) cautions that these running events need to be creative and to differentiate themselves from other running events to stay competitive. Harada (2016) adds that marathon events are in the so-called maturity stage of the product life cycle, and, therefore, the runners are becoming selective when choosing the most desired running event.

In fact, not all running events have been economically successful. For example, Masubuchi (2013) reports that the Kyoto Marathon lost 230 million yen in 2012 due to increased traffic security measures. In 2018, the Tanegashima Marathon, with 30 years of history behind it, decided to discontinue the event due to increased competition. Likewise, the Izu Marathon and the Yokohama Women's Marathon were terminated in 2017 and 2014, respectively (Iijima, 2017). Just as in the aforementioned cases, sporting event managers need to focus on ways to increase revenue and make the running events profitable. Moreover, for an event to thrive for a sustained period of time, it is necessary to build a strong brand image for the event as well as to develop the event's brand equity (Keller, 2013).

Numerous studies have focused on branding sporting events from various perspectives. Previous research has investigated fans' brand associations or any memories that the fans link to a specific brand, for an intercollegiate athletic team (Ross et al., 2009), for a professional sports league (Kunkel et al., 2014), and for professional sports teams (Bauer et al., 2005, 2008; Gladden and Funk, 2002; Kunkel et al., 2016; Ross, 2006, 2007; Ross et al., 2006; Ross et al., 2007; Ross et al., 2008) using Keller's (1993) consumer-based brand equity framework. In the sport tourism literature, sporting events' brand images were measured using the destination image [e.g., Kaplanidou, 2009; 2010; Papadimitriou et al., 2016)], affective brand image (Kaplanidou and Vogt, 2007), and attribution of meanings (Kaplanidou and Vogt, 2010) frameworks. Even though scholars have studied the sports brand association from different perspectives and in different contexts, participant sporting events have received little attention. Moreover, sports tourism literature has not explored the sporting event brand image using Keller's (1993) framework. Therefore, the current study attempts to fill the gap in the literature by adopting the aforementioned framework in the participant-sport setting. Additionally, the study intends to understand the effects of an event's brand equity on the host city's brand equity. The current study focused on the Okinawa Marathon, a participant-based sporting event with approximately 16,000 runners, held each year in Okinawa, Japan. The study contributes to scholarship by the understanding of brand equity in the participant sport setting and brand equity's influence on place marketing.

Literature review

Events like festivals, MICE events, or sports competitions are occurrences that gain rising attention in science (see: Cudny, 2016; Getz, 2008; Getz and Page, 2016). They generally may be divided into planned and unplanned. The event analysed in this study falls into the category of planned events which

> by definition, have a beginning and an end. They are temporal phenomena, and with planned events, the event programme or schedule is generally planned in detail and well-publicized in advance. Planned events are also usually confined to particular places, although the space involved might be a specific facility, a very large open space, or many locations simultaneously or in sequence.
>
> (Getz and Page, 2016, p. 46)

According to Getz (2008, p. 404), events can be divided into different categories according to their scale and type. In the latter division, there are the following groups of events: cultural celebrations (e.g., festivals), political and state (like VIP visits, royal occasions), arts and entertainment (e.g., concerts), business and trade (meetings, conventions, trade shows, markets), educational and scientific (like seminars), sports competitions (amateur/ professional; spectator/participants); recreational (sports or games for fun), and private events (weddings, parties) (Getz, 2008, p. 404). The Okinawa Marathon characterised in our chapter falls into the category of amateur participant sports competitions.

The organisation of events has different impacts on society, economy, culture, and host cities (Cudny, 2016). Events generate event tourism, booster the economic development due to investments and tourism revenues; they are an interesting free time offer for inhabitants. Events create social opportunities, strengthen integration (e.g., multicultural festivals), and develop social capital (see: Arcodia and Whitford, 2007; Cudny, 2013). When speaking of events impacts, brand creation and promotion of host areas is one of the key issues (Cudny, 2020).

As Kotler and Armstrong (2010, p.255) stated, brand is "a name, term, sign, symbol, design, or a combination of these that identifies the products or services of one seller or group of sellers and differentiates them from those of competitors". When a brand is applied to a place (a country, region or city), we can talk about place brand or city brand. The process of branding in relation to places is most often perceived as the long-term development programme encompassing the mixture of development of place offer (so-called place product) for tourists, entrepreneurs, inhabitants (see: Anholt, 2008; Lucarelli, 2018) and place promotion and marketing communication (see: Cudny, 2019). Branding is part of place marketing which encompasses promotion, building place brand, and redefining image in order to successfully attract residents, firms, and tourists and sell the place products. When events are involved in place marketing, we may talk about the concept of place event marketing. The concept is twofold and it includes (1) creation of events as attractions enriching the place product and (2) city promotion and brand creation through brand equity evoked by events (Cudny, 2019).

Over the past two decades, sports organisations have focused on long-term strategic brand management to increase brand value over time rather than focusing on short-term profits (e.g., Gladden et al., 2001; Ross, 2006). For example, sports teams have been hiring branding agencies to create brand logos that represent their teams' images. Having strong and favourable brand equity allows sports organisations to attract and retain consumers, to charge premium pricing, and to protect themselves from detrimental financial losses (e.g., Gladden and Funk, 2001; Kunkel et al., 2016; Ross, 2006). For fans, favourable brand equity allows for easier information processing, provides confidence in pre-and post-purchase decisions, and improves user satisfaction (Ross, 2006). It also creates loyal fans that will stick with and support a sports organisation even through continuous losing seasons and despite a lack of championships. The question, then, becomes this: How can sports marketers build strong brand equity?

Success on the field is usually and myopically deemed to be the only important factor that contributes to sports brand equity. However, numerous different contexts exist in sports, and various other factors affect brand equity. First of all, success on the field is not a given for any formidable team or athlete. Upsets are prone to happen, and only one team is crowned champions at the end of the season. Additionally, in participant-sport settings such as a road race, winning is considered less important than other factors, such as a sense of accomplishment or completing the race with friends and family.

The literature on sport brand equity has evolved around Keller's (1993) consumer-based brand equity. Keller proposed that brand knowledge, a multidimensional construct, is what ultimately decides the strength of brand equity. Within brand knowledge, two important dimensions exist, namely brand awareness and brand image. Brand awareness is the consumers' ability to remember and identify a brand from memory, while brand image refers to the "perceptions about a brand as reflected by the brand associations held in consumer memory" (Keller, 1993, p.3). Therefore, individuals' brand associations—thoughts, feelings, and ideas about a brand in memory—become the foundation for building strong brands (Aaker, 1991). Moreover, the type, uniqueness, the degree of favourability, and the strength of brand associations will determine the brand's success in the market (Keller, 2013).

Keller (1993) suggested that brand associations are comprised of attributes, benefits, and attitudes. Attributes are features of a good or a service that can be further divided into product-related and non-product-related attributes. Product-related attributes are those factors that are needed to perform the function(s) expected by consumers. Non-product-related attributes, on the other hand, are the external conditions of a good or a service that do not impact the overall performance of the good or the service, such as price information, packaging or product appearance information, user imagery, and usage imagery. Next, benefits are consumers' personal values they attach to a good or a service and these benefits can be further categorised into functional, experiential, and symbolic benefits (Park et al., 1986). Lastly, attitudes are consumers' overall evaluations of a brand and are usually formed by considering the overall assessment of a brand's uniqueness, strength, and the favourability of its attributes and benefits (Keller, 1993).

Within sport settings, most of the research has been aimed at understanding the factors that underly brand associations. Using Keller's (1993) conceptual framework, Gladden and Funk (2002) developed the team association model (TAM), in which they identified 16 dimensions of brand associations. In their study, the attributes comprised eight dimensions (success, head coach, star player, management, stadium, logo design, product delivery, and tradition), the benefits had five dimensions (identification, nostalgia, pride in place, escape, and peer group acceptance), and the attitude was composed of three dimensions (importance, knowledge, and affect). In the sport management literature, the TAM has since served as a starting point to understand brand associations in team sports settings. Bauer et al. (2005) later attempted to extend the TAM in the professional German soccer setting. They highlighted the adequacy of the model in team sports as well as its importance for a team's economic success. Ross et al. (2006) developed the team brand association scale (TBAS), identifying 11 brand associations that consumers linked to teams (nonplayer personnel, team success, team history, stadium community, team play, brand mark, commitment, organisational attributes, rivalry, concessions, and social interaction) and pointed out the significance of the scale to sport managers in creating favourable associations to attract and retain consumers. Unlike the previous studies, which approached brand associations from a researcher's perspective, Ross et al.'s (2006) approach was based on a fan's perspective. The TBAS was later applied to different sport contexts (Ross et al., 2007) as well as in other studies that used the scale to understand the relationships between brand association and other constructs (Ross, 2007; Ross et al., 2008, 2009). Bauer et al. (2008) extended the research of Gladden and Funk (2002) by looking at the causal relationships between attributes, benefits, and attitudes. They found that both the product- and the non-product-related attributes affected benefits, while the benefits, in turn, influenced fans' attitudes. They also contributed to the literature by measuring uniqueness, strength, and favourability, a research limitation mentioned in the previous literature. Kunkel et al. (2014) later expanded the field by examining consumer-based league brand associations. They constructed a single-item scale, adding six new associations to the league brand association scale.

Ross (2006) took a different approach to understand brand equity. He argued that previous studies took a product-oriented approach, whereas team sports are more service-oriented and thus require an alternative approach. Berry's (2002) service brand equity model served as a foundation for Ross' (2006) model, according to which antecedents to brand equity should also be incorporated in the brand equity model (organisation-induced, market-induced, and experience-induced antecedents) to understand service brand equity. Organisation-induced antecedents refer to the marketing strategies and activities of the team, while market-induced antecedents are sources for consumers to obtain brand information that did not originate from the brand, namely word-of-mouth communications and publicity. Lastly, experience-induced antecedents are actual service experiences encountered by the fans. Ross (2006) proposed that these three antecedents build brand equity and eventually produce different outcomes.

While Keller's (1993) framework for brand association research has been applied to professional and intercollegiate teams, participant sporting events have received considerably less attention. In the sport tourism literature, research related to the brand image of participant-sport events used the destination image framework. A destination image is "the sum of beliefs, ideas, and impressions that a person has of a destination" (Crompton, 1979, p.18). The framework consists of cognitive, affective, and conative characteristics that can have functional, psychological, and unique components, which can be further used to position a destination relative to its competitors (Echtner and Ritchie, 1993). Cognitive characteristics relate to knowledge, while affective refers to the feelings held by individuals. Conative characteristics include the individual's intent to carry out the behaviour (Pike and Ryan, 2004). Kaplanidou (2010) investigated marathon runners in Athens, Greece. Using qualitative analysis, the study found six image-related themes pertaining to the event and the host city: historical, emotional, organisational, physical, environmental, and social. Similarly, Kaplanidou and Vogt (2010) attempted to understand the meanings that sport event tourists attach to small-scale recurring sport event experiences. They identified the following five meanings: organisational, environmental, physical activity, social, and emotional. The organisational meaning refers to the event organiser's service offerings. The environmental meaning has to do with the scenery of the event and the host city. The physical activity relates to the benefits received from participation, and social meaning refers to the opportunities to socialise with other participants. Lastly, the emotional relates to the fun, the excitement, and the enjoyment felt during participation. Papadimitriou et al. (2016) investigated the Summer Universiade event and its event images. They uncovered additional brand image dimensions, notably the competition, the cultural dimensions, as well as unclassified aspects.

Furthermore, they emphasised that unique event characteristics create additional brand images.

Kaplanidou and Vogt (2007) examined the effects of small-scale sporting events on a community. The authors found that the sports event's image positively affected the destination image, underlining the power of events as formations agents of the destination image. Other studies (e.g., Hallmann and Breuer, 2010a, b) have found a similar effect by the event image on other constructs such as revisit intentions and further tourism development.

The current study first aims to identify event brand associations in a participant sport setting using Keller's (1993) framework. Furthermore, following Breuer et al.'s (2008) findings, the study will investigate the causal effects of attributes, benefits, and attitudes. Lastly, the study will determine the effects of an event's brand association on the event's brand equity and the host place's brand equity.

Study site, the event, and methods

Study site

Okinawa, the southernmost prefecture in Japan, consists of over 100 islands in the subtropics. It is also one of the most visited prefectures in Japan because of its nature and scenery as well as its unique history, food, sports, and culture (Okinawa prefecture, 2019). Okinawa usually ranks among the top three prefectures in the best domestic travel destinations ranking by travellers (Hayashi, 2018). Okinawa also outranks other prefectures in the overall travel satisfaction ranking (Trip Editor, 2018). Its rivals are usually Hokkaido and Kyoto (for additional information on Okinawa prefecture, please refer to the following website: https://www.pref.okinawa.jp/site/chijiko/kohokoryu/foreign/english/index.html).

The event

The Okinawa Marathon is usually held on the third Sunday of February. The event started in 1955 as the Shimpo[1] Naha Marathon and moved to central Okinawa in 1993. The Okinawa Marathon has offered different types of races throughout its history. Among the different types, the full marathon and the 10-km run are the main two races of the event. The full marathon course spans five municipalities (including several towns): Chatan, Kadena, Kitanakagusuku, Okinawa, and Uruma. Furthermore, the course includes a short route into the U.S. Kadena Air Base (Figure 5.1.), a unique experience for most runners as access is usually prohibited (for information on the course route, please see: https://okinawa-marathon.com/eng/course_map.html). The event attracts participants from all over Japan and even from overseas. However, the majority of the runners in recent years have been Okinawa residents. The limit on the number of runners is 16,500 (13,500 for the full marathon and 3,000 for the 10-km run).

An association comprised of nine municipalities (Local Government Association of Central Okinawa) form a committee (Okinawa Marathon Executive Committee) to organise the event along with Okinawa Field Athletic

Figure 5.1 Runners going into Kadena Airbase. Source: Authors.

Society, TrusTec Mizuno Inc. (Designated Manager of Okinawa Comprehensive Park), and two media companies, namely Ryukyu Shimpo (newspaper) and Okinawa Television. For the promotion of the event, the organisers use flyers, direct mail, Internet ads, Social Networking Systems (Facebook and Twitter), homepage, and booths at marathon expos and tourism expos as well as disseminating information through the Okinawan prefectural tourism agencies abroad. The goal of their strategy is to retain the current customers and to attract new ones by increasing the awareness of the event by distributing flyers to current marathon runners and encouraging early registration. For customer retention, the organisers send direct mails to past event participants (e.g., 2,550 mails in 2017) in September, five months ahead of the event (N. Okino, personal communication, June 5, 2017). For new customer acquisitions, they travel extensively outside of Okinawa to other parts of Japan and abroad to attract visitors. For example, in 2016, the event organisers travelled from July to December to eight domestic and two international marathon events and expos to promote the upcoming event. At the Hokkaido Marathon Expo 2016, event organisers promoted the event along with other marathon events held in Okinawa. Two thousand leaflets promoting the upcoming Okinawa Marathon were distributed at this expo and 17 guests signed up on-site for the event (Okinawa prefecture, 2017). Furthermore, at international tourism expos in Singapore and Taipei, the organisers held seminars and sold travel packages to Okinawa. The Singapore event garnered 50 guests to the seminar, of which 6 purchased Okinawa Marathon packaged tours on-site and an additional 12 eventually ran the event (Okinawa prefecture, 2017).

Due to these efforts, participation from outside of Okinawa is growing; however, the majority of the participants in 2017 reside in the Okinawa prefecture

(73.26% or 10,965 runners). Participants from outside of the Okinawa prefecture comprise 18.81% (2,816 runners) of the total number of participants, and overseas participants make up 7.92% (1,186 runners) of the total runners. These participant numbers from outside of the Okinawa prefecture and from overseas are 1.5 times and 3.7 times greater than those in 2011 (N. Okino, personal communication, February 3, 2018). The total number of participants has been flat from 14,967 runners in 2017, 14,768 runners in 2018, 15,418 runners in 2019, and 14,627 runners in 2020 (Okinawa Marathon, 2020).

Study methods

To understand the brand associations of runners at the Okinawa Marathon, one of the authors conducted a direct observation of the event by participating and completing the full marathon a year before data collection. The following year, we conducted a free-thought listing procedure to solicit responses from other runners. With permission from the event organisers, thought-listing online form links were posted on the official Okinawa Marathon's Twitter and Facebook accounts. Respondents ($N = 71$) were asked to list as many thoughts as they could using an Internet website. Of the 71 administered responses, 12 were not completed; therefore, they were excluded from further analyses. Griffin and Hauser (1993) recommended having about 20–30 respondents to capture 90% of the responses. The number of completed responses was 59 and well above the suggested number and therefore was deemed appropriate. The researcher and an outside coder conducted a content analysis of the listed thoughts to identify broad categories for the scale. Each individual was responsible for analysing and selecting an appropriate category.

Respondents were mainly male (73%), office workers (69%), Okinawa residents (51%), and had participated in the Okinawa Marathon in the past (70%). The average age was 43.33 years ($SD = 10.86$ years). The respondents provided a total of 191 individual thoughts regarding the Okinawa Marathon. Next, these thoughts were cross-examined by two researchers to identify common themes and to categorise them into groups. Once the procedure was completed, multiple event organisers were asked to review the items for face validity and content validity. Minor changes were made to the items as a result.

The categories included product-related attributes (course-related: difficulty, sightseeing value, city view, beautiful scenery, and the U.S. base; runners-related: runners in costumes, invited runners, and other runners; others: time limit) and non-product-related attributes (other attractions at the event, sponsors, support on the streets, volunteers, registration, access to the event, obtaining bibs, participation prize, logos and mascots, event history, theme song, and cultural events). Attributes were measured on a 7-point semantic differential scale for their uniqueness, strength of association, and favourability. Benefits were also identified in the free-thought listing process. Furthermore, items from Masters et al.'s (1993) study on the marathon runner's motives were added to the questionnaire. Benefits were measured for their uniqueness and strength of association as Bauer et al.

(2008) argued that benefits are favourable in themselves. Attitudes were measured using Cunningham and Kwon's (2003) four-item semantic differential scale ranging from one to seven (unpleasant–pleasant, boring–exciting, dull–entertaining, worthless–valuable). Similar to Bauer et al.'s (2008) study, an aggregate variable was constructed by averaging the uniqueness, strength, and favourability of the variables. Lastly, brand equity for the event and the place were measured using Yoshida et al.'s (2013) two items for capturing the cognitive and affective dimensions.

Exploratory factor analysis

The researcher constructed a questionnaire regarding product-related and non-product-related attributes, benefits, attitudes, and brand equity items. The questionnaire was administered to participants at the 2017 Okinawa Marathon ($N =$ 462). Five university students were hired and trained to collect data at the event site. They were instructed to approach runners after the completion of the run. Of the total respondents, 109 respondents ran the half-marathon or the 10-km run, and additional 3 respondents were at the event to cheer for the runners. Additionally, 46 respondents did not indicate their involvement. Thus, the number used for the analysis was 304. Respondents were mainly male (79.3%), single (64.2%), resided in Okinawa (81.9%), and had higher than a college degree (53.9%). Their average age was 32.02 ($SD = 10.05$).

Exploratory factor analysis (EFA) with maximum likelihood extraction and principal axis rotation was conducted on brand association and benefits items to understand the underlying dimensions of the scale. The researchers used eigenvalue equal to or greater than 1 as the initial criterion and a scree plot to determine the number of factors to retain. The results indicated extracting eight factors, which explained 64.76% of the total variance. Items with factor loading values over 0.4 were considered in the interpretation of the factors. The researchers named these eight factors as follows: attributes (runners, course, uniqueness of the event, peripheral event attributes, and operation) and benefits (psychological, social, and health). No distinctions were made between product-related and non-product-related attributes as EFA's results indicated that the uniqueness of the event included both. Cronbach's alpha for these eight factors ranged from .64 to .88, indicating adequate internal consistency.

Results

Descriptive statistics, factor loadings, and the Cronbach's alpha coefficient for the brand association items are shown in Table 5.1. To test the causal relationships, the researcher conducted structural equation modelling (SEM) using AMOS 25 (Figure 5.2). Attributes and benefits were treated as second-order latent variables. The researcher used the following indices to assess the fit of the model to the data: the chi-square goodness-of-fit statistic (x^2), relative chi-square (x^2/df), the comparative fit index (CFI), the Tucker-Lewis index (TLI), and the root mean square error

Table 5.1 Descriptive statistics, factor loadings and alpha coefficients

Variable	M	SD	Factor loadings	Alpha coefficients
ATTRIBUTES				
Runners				0.83
Costumes	4.77	1.38	0.778	
Invited runners	3.83	1.39	0.722	
Other runners	4.14	1.26	0.635	
Time limit	4.12	1.21	0.497	
Course				0.78
Scenery	4.68	1.22	0.872	
City view	4.44	1.30	0.815	
Sightseeing	4.57	1.41	0.569	
Uniqueness of event				0.64
Volunteers	5.62	1.25	0.724	
Support on streets	5.84	1.22	0.625	
US base	6.33	0.93	0.469	
Peripheral event attributes				0.88
Logo & mascot	4.49	1.36	0.791	
Sponsors	4.47	1.28	0.692	
Event history	4.86	1.40	0.690	
Cultural events	4.91	1.33	0.632	
Participation prize	4.57	1.14	0.562	
Management				0.80
Obtaining bibs	4.00	1.26	0.666	
Access to events	4.00	1.28	0.634	
Registration methods	3.98	1.18	0.603	
BENEFITS				
Psychological				0.80
Self-esteem	4.66	1.52	0.963	
Meaning in life	4.35	1.63	0.897	
Personal goal achievement	4.87	1.63	0.486	
Community identification	4.47	1.62	0.469	
Psychological coping	3.84	1.57	0.442	
Social				0.84
Feeling close to others	4.63	1.53	0.994	
Interaction with others	4.31	1.54	0.730	
Health				0.69
Control weight	4.23	1.60	0.790	
Improve health	4.61	1.51	0.612	
ATTITUDES				0.86
Pleasant	6.05	1.26		
Exciting	5.75	1.53		
Entertaining	5.89	1.45		
Valuable	5.93	1.40		

Source: Authors.

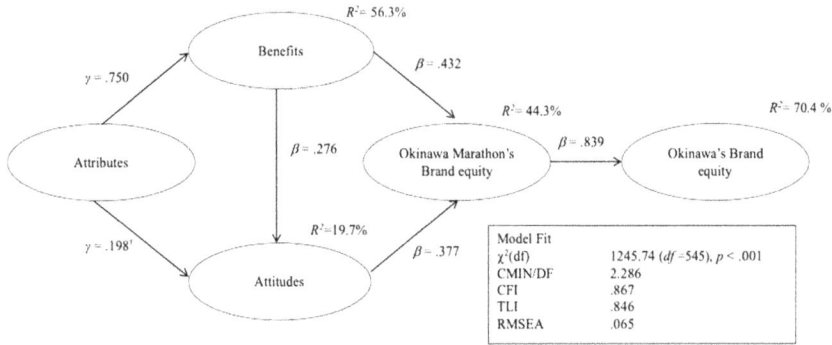

Figure 5.2 Results of the model fit, path coefficients, and effect sizes. Note: † denotes significance at *p* <.10 level; all others are significant at *p* <.05 level. Source: Authors.

of approximation (RMSEA). The general criteria for these indices were the following: the chi-square should not be significant, the relative chi-square (CMIN/DF) should be below three, the CFI and the TLI should be greater than .90, and the RMSEA should be below .05 to be considered to have an excellent fit, below .08 to have fair fit, below .10 to have mediocre fit, and anything above .10 to have a poor fit (Browne and Cudeck, 1993; MacCallum, Browne, and Sugawara, 1996).

Results of the SEM indicated a mediocre fit to the data: $x^2 = 1245.74$ ($df = 545$), $p < .001$, CMIN/DF = 2.29; CFI = .87; TLI = .85; RMSEA = .065. Further analysis indicated that the attributes significantly and positively affected benefits ($\gamma = .750$, $p < .001$). Attributes also marginally influenced attitudes ($\gamma = .198$, $p = .088$); however, attributes had no effect on the Okinawa Marathon's brand equity ($\gamma = -.025$, $p = .820$). Benefits significantly and positively affected both attitudes ($\beta = .276$, $p = .029$) and the Okinawa Marathon's brand equity ($\beta = .432$, $p = .001$). Likewise, attitudes significantly and positively influenced the Okinawa Marathon's brand equity ($\beta = .377$, $p < .001$). Lastly, the Okinawa Marathon's brand equity had a significant and positive impact on Okinawa's (place) brand equity ($\beta = .839$, $p < .001$). The model explained 56.3% of the variance in benefits, 19.7% of the variance in attitudes, 44.3% of the variance in the Okinawa Marathon's brand equity, and 70.4% of the variance in Okinawa's brand equity. Figure 2 depicts the relationship of the constructs in the study.

Discussion

The purpose of the study was to first explore the brand association dimensions in the participant sport setting. Next, the study aimed to understand the causal relationships between attributes, benefits, and attitudes. Lastly, the effects of brand association were tested on an event's brand equity as well as on brand equity of the host place. Theoretical contributions and practical implications will be discussed in this section.

The results suggested that the event's brand association comprised eight unique factors (five attributes and three benefits) of a marathon event. While some of the factors (e.g., the organisational and the social) were similar to those of previous studies (Kaplanidou, 2009, 2010; Kaplanidou and Vogt, 2010; Papadimitriou et al., 2016), the current study makes original contributions to the literature. Some of them will be discussed.

The uniqueness of the event emerged as one of the important attributes, comprising both product-related (U.S. base) and non-product-related attributes (street support and volunteers). The reason why both of them loaded on one factor was probably that the runners have gotten quite accustomed to various forms of help and support from the volunteers and the locals on the streets of the Okinawan road races. In fact, Matsuyama (2014) argues that the road races in Okinawa are among the top ones in terms of the number and the variety of street supporters. Respondents might have considered these unique non-product-related attributes as an essential part of the race and, thus, deemed them to be part of the product. In relating the findings to practice, even though the course runs through a U.S. base, the organisers failed to take advantage of this competitive edge in their flyers, pamphlets, and web sites. They should take full advantage of this opportunity in the future.

The presence of other runners was considered to be an important attribute that eventually led to the event's brand equity. Invited runners, runners in costumes, and the mere presence of other runners added to the evaluation of the event. The Okinawa Marathon organisers understand the importance of these runners and have numerous pictures of these individuals in their marketing collateral. Next, the reason why time limits loaded on the same runners' factor is probably because of the drama that it creates. Usually, there are multiple checkpoints in a road race with various time limits depending on the level of the race. In the case of the Okinawa Marathon, there are seven checkpoints, with the last one being the entrance gate into the stadium (400 meters before the finish line, 6 hr and 15 min time limit) (Figure 5.3). Runners who cannot enter the stadium within the specified time limit are required to quit the race (i.e., are disqualified from the event). This aspect creates a challenge and a drama for the novice runners who are on the verge of the time limit. We believe the time limit was grouped with the runners' factor for this reason.

Attributes related to the course, peripheral events, and event operations emerged to be important in assessing brand association. These factors were similar to the findings from previous studies, and information related to these aspects is abundant in the marketing materials at the Okinawa Marathon. Concerning the benefits of brand association, previous studies have not distinguished them from attributes. The current study identified psychological (self-esteem, personal goal achievement, psychological coping, meaning in life, and community identification), social (feeling connected and interacting with others), and health (health improvement, and weight control) benefits. While the

Figure 5.3 Runners going into the stadium. Source: Authors.

current study failed to categ orise these into Keller's (1993) framework (functional, emotional, and symbolic benefits), the findings represent benefits from tourism and leisure areas (e.g., Masters et al., 1993; Papadimitriou et al., 2016). These benefits are well represented in the flyers and pamphlets, with pictures showcasing runners having fun with other runners and depicting the runners expressing a sense of accomplishment. Besides, the event organisers held two marathon running clinics outside of Okinawa during the summer. These events allowed runners to set their personal goals and practices and to achieve them at the time of the event.

The current study also focused on the causal relationships of attributes, benefits, and attitudes. Results confirmed the relationship presented by Bauer et al. (2008) in the participant-sport setting. Additionally, the study focused on how the event's brand association influenced the event's overall brand equity and eventually Okinawa's brand equity. The results confirm this relationship. This is an important contribution to the literature, confirming Keller's (1993) model in the participant sport context and expanding the sports marketing and place marketing fields. Sport managers need to focus on incorporating the destination's image into the event as previous studies have identified a positive causal relationship between the two (e.g., Florek and Insch, 2011; Frost, 2008; Hallmann et al., 2010; Xing and Chalip, 2006). Likewise, previous studies have found an event's brand image to positively influence or revisit intentions to the destination (Hallmann and Breuer, 2010a; Xing and Chalip, 2006). Correspondingly, sports organisers for the Okinawa Marathon have usually included pictures of the blue sky or the ocean, which are typical brand associations of Okinawa.

Limitations and future research

There are several limitations to the study that need to be addressed as the study was exploratory in nature. First, the current study only solicited responses from the participants of the event. It seems quite important that future research should investigate the perceptions of future potential participants. This would allow a better understanding of the brand association and its effects on the brand equity of both the event and the place. Second, most of the current respondents were Okinawa residents. While these people are important in forming perceptions of travel destinations, future research should explore the different brand associations that participants outside of the place may hold. Third, it might be equally interesting to examine differences between first-time participants and repeat participants. Additionally, different levels of running involvement may introduce different perspectives to the meanings attached to the event. The findings from these different segments will allow managers to effectively segment and target future runners. Fourth, the items of the study need more sophistication in different contexts. The applicability of the scale is unknown, and further testing of the scale is warranted. This would provide deeper an understanding of brand association in various participant sport settings. Fifth, the study did not include the brand awareness factor, which is an important part of brand knowledge in Keller's (1993) framework. Future research should address this shortcoming. Sixth, each brand association was measured using a single item. While Rossiter (2002) contends that single-item scales are adequate for measuring core marketing concepts, future research should employ multiple items to tap into the brand association. Lastly, brand association and the brand equity of events should be tested on other outcome constructs, such as future intentions to participate in the event. Understanding these relationships will contribute to scholarship and should provide important practical implications.

Conclusion

In summary, this study focused on how brand associations affect brand equity in the context of an amateur participant sport. The study offered partial support for Keller's (1993) brand association model. The findings confirmed an eight-factor model (five attributes and three benefits) which leads to the formation of attitudes in the participant sport setting, confirming Bauer et al.'s (2008) study. Lastly, the results revealed that an event's brand equity influenced the place's brand equity, which is an important contribution to existing scholarship. The results of the study should assist event managers in attracting new participants and retaining the existing runners through the effective use of attributes and benefit, which is crucial in their event marketing strategies (Cudny, 2020).

Note

1   Shimpo refers to *Ryukyu Shimpo*, the oldest newspaper in Okinawa.

References

Aaker, D.A. (1991). *Managing Brand Equity*. New York: The Free Press.

Anholt, S. (2008). Place branding: Is it marketing, or isn't it?, *Place Branding and Public Diplomacy*, *4*(1), 1–6.

Arcodia, C., & Whitford, M. (2007). Festival attendance and the development of social capital. *Journal of Convention & Event Tourism*, *8*(2), 1–18.

Bauer, H.H., Sauer, N.E., & Schmitt, P. (2005). Customer-based brand equity in the team sport industry: Operationalization and impact on the economic success of sport teams. *European Journal of Marketing*, *39*(5/6), 496–513.

Bauer, H.H., Stokburger-Sauer, N.E., & Exler, S. (2008). Brand image and fan loyalty in professional team sport: A refined model and empirical assessment. *Journal of Sport Management*, *22*, 205–226.

Berry, L. (2002). Cultivating service brand equity. *Journal of the Academy of Marketing Science*, *28*, 128–137.

Browne, M.W., & Cudeck, R. (1993). Alternative ways of assessing model fit. In K.A. Bollen & J.S. Long (Eds.), *Testing Structural Equation Models* (pp. 136–162). Newbury Park, CA: Sage.

Crompton, J.L. (1979). An assessment of the image of Mexico as a vacation destination and the influence of geographical location upon that image. *Journal of Travel Research*, *17*(4), 18–23.

Cudny, W. (2013). Festival tourism–the concept, key functions and dysfunctions in the context of tourism geography studies. *Geographical Journal*, *65*(2), 105–118.

Cudny, W. (2016). *Festivalisation of Urban Spaces: Factors, Processes and Effects*. Springer, Cham.

Cudny, W. (2019). *City Branding and Promotion: The Strategic Approach*. London: Routledge

Cudny, W. (2020). The concept of place event marketing: Setting the agenda. In Cudny, W. ed. *Urban Events, Place Branding and Promotion: Place Event Marketing*, pp. 1–25. London: Routledge.

Cunningham, G.B., & Kwon, H. (2003). The theory of planned behavior and intentions to attend a sport event. *Sport Management Review*, *6*, 127–145.

Echtner, C.M., & Ritchie, B.J.R. (1993). The measurement of destination image: An empirical assessment. *Journal of Travel Research*, *31*(4), 3–13

Florek, M. & Insch, A. (2011). When fit matters: Leveraging destination and event image congruence. *Journal of Hospitality Marketing & Management*, *20*(3–4), 265–286.

Frost, W. (2008). Projecting an image: Film-induced festivals in the American west. *Event Management*, *12*, 95–103.

Getz, D. (2008). Event tourism: Definition, evolution, and research. *Tourism Management*, *29*(3), 403–428.

Getz, D., & Page, S. (2016). *Event Studies: Theory, Research and Policy for Planned Events*. London: Routledge.

Gladden, J.M., & Funk, D.C. (2001). Understanding brand loyalty in professional sport: Examining the link between brand associations and brand loyalty. *International Journal Sports Marketing & Sponsorship*, *3*(1), 54–81.

Gladden, J.M., & Funk, D.C. (2002). Developing an understanding of brand associations in team sport: Empirical evidence from consumers of professional sport. *Journal of Sport Management*, *16*, 54–81.

Gladden, J.M., Irwin, R.L., & Sutton, D.C. (2001). Managing North American major professional sport teams in the new millennium: A focus on building brand equity. *Journal of Sport Management*, *15*(4), 297–317.

Griffin, A., & Hauser, J.R. (1993). The voice of the customer. *Marketing Science*, *12*(1), 1–27.

Hallmann, K., & Breuer, C. (2010a). Image fit between sport events and their hosting destinations from an active sport tourist perspective and its impact on future behavior. *Journal of Sport & Tourism*, *15*(3), 215–237.

Hallmann, K., & Breuer, C. (2010b). The impact of image congruence between sport event and destination on behavioral intentions. *Tourism Review*, *65*(1), 66–74.

Hallmann, K., Kaplanidou, K., & Breuer, C. (2010). Event image perceptions among active and passive sport tourists at marathon races. *International Journal of Sports Marketing & Sponsorship*, 37–52.

Harada, M. (2016). Houwa joutai no shimin marason… Taikai ikinokori no kagi ha? [Saturated marathon market: The key to survival]. *Yomiuri Shimbun*. http://www.yomiuri.co.jp/fukayomi/ichiran/20151225-OYT8T50120.html?page_no=1

Hayashi, K. (2018). Kankou de ikitai todoufuken & shichouson rankingu 2018 kanzenban [Top 10 travel destinations in Japan. Full version]. *Diamond* Online. https://diamond.jp/articles/-/187904.

Iijima, K. (2017). Rannaa okizari shimin marason, baburu houkai. [Runners left stranded. Burst of the marathon event bubble]. https://www.nikkei.com/article/DGXMZO12775770Q7A210C1000000/

Inoue, T. (2018). Nihon kakuchi no marason taikai ga gaikokujin rannaa kangei no riyuu. [Reasons why marathon events in Japan welcome foreign runners]. https://www.newsweekjapan.jp/nippon/season2/2018/04/212266.php

Kaplanidou, K. (2009). Relationships among behavioral intentions, cognitive event and destination images among different geographic regions of Olympic Games spectators. *Journal of Sport & Tourism*, *14*(4), 249–272.

Kaplanidou, K. (2010). Active sport tourists: Sport event image considerations. *Tourism Analysis*, *15*, 381–386.

Kaplanidou, K., & Vogt, C. (2007). The interrelationship between sport event and destination image and sport tourists' behaviours. *Journal of Sport & Tourism*, *12*(3–4), 183–206.

Kaplanidou, K., & Vogt, C. (2010). The meaning and measurement of a sport event experience among active sport tourists. *Journal of Sport Management*, *24*(5), 544–566.

Keller, K.L. (1993). Conceptualizing, measuring, and managing customer-based brand equity. *Journal of Marketing*, *57*, 1–22.

Keller, K.L. (2013). *Strategic Brand Management: Building, Measuring, and Managing Brand Equity* (4th ed.). New York: Pearson.

Kotler, P., & Armstrong, G. (2010). *Principles of Marketing*. Upper Saddle River: Pearson Education.

Kunkel, T., Funk, D.C., & King, C. (2014). Developing a conceptual understanding of consumer-based league brand associations. *Journal of Sport Management*, *28*, 49–67.

Kunkel, T., Doyle, J.P., Funk, D.C., Du, J., & McDonald, H. (2016). The development and change of brand associations and their influence on team loyalty over time. *Journal of Sport Management*, *30*, 117–134.

Lucarelli, A. (2018). Co-branding public place brands: Towards an alternative approach to place branding. *Place Branding and Public Diplomacy*, *14*(4), 260–271.

MacCallum, R.C., Browne, M.W., & Sugawara, H.M. (1996). Power analysis and determination of sample size for covariance structure modeling. *Psychological Methods, 1*, 130–149.

Masters, K. S., Ogles, B. M., & Jolton, J. A. (1993). The development of an instrument to measure motivation for marathon running: The motivations of marathoners scales (MOMS). *Research Quarterly for Exercise and Sport, 64*(2), 134–143.

Masubuchi (2013). Sankahi ichi manen demo akaji? Marason taikai no daidokoro jijou. [Marathon events financial status: Are they losing money with 10,000 yen participation fees?] https://www.nikkei.com/article/DGXNASDJ2701D_X21C12A2000000/

Matsuyama, K. (2014, Autumn). Omotenashi ha sekai ichi!? NAHA marason tettei kaibou [Hospitality is top in the world: In-depth analysis of NAHA marathon. *Number Do*, 48–51.

Okinawa Marathon (2020). Okinawa Marathon. https://okinawa-marathon.com/

Okinawa Prefecture (2017). *Heisei 28 nendo supotu tsuurizumu senryaku suishin jigyou jisshi houkokusho.* [2016 Sport Tourism Strategy Promotin Project Report]. Okinawa prefectural government. https://www.pref.okinawa.lg.jp/site/bunka-sports/sports/documents/h28yuukyaku3.pdf

Okinawa Prefecture (2019). *Heisei 30 nen kankou toukei jittai chousa.* [Statistics on Okinawa tourism in 2018] Okinawa Prefectural Government. https://www.pref.okinawa.jp/site/bunka-sports/kankoseisaku/kikaku/report/tourism_statistic_report/documents/3_mannzokudobunseki_1.pdf

Papadimitriou, D., Apostolopoulou, A. & Kaplanidou, K. (2016). Participant-based brand image perceptions of international sport events: The case of the Universiade, *Journal of Convention & Event Tourism, 17*(1), 1–20.

Park, W., Jaworski, B.J., & MacInnis, D.J. (1986). Strategic brand concept-image management. *Journal of Marketing, 50*(4), 135–145.

Pike, S. & Ryan, C. (2004). Destination positioning analysis through a comparison of cognitive, affective, and conative perceptions. *Journal of Travel Research, 42*, 333–342. https://doi.org/10.1177/0047287504263029

Ross, S.D. (2006). A conceptual framework for understanding spectator-based brand equity. *Journal of Sport Management, 20*(1), 22–38.

Ross, S.D. (2007). Segmenting sport fans using brand associations: A cluster analysis. *Sport Marketing Quarterly, 16*, 15–24.

Ross, S.D., James, J.D., & Vargas, P. (2006). Development of a scale to measure team brand associations in professional sport. *Journal of Sport Management, 20*, 260–279.

Ross, S.D., Bang, H., & Lee, S. (2007). Assessing brand associations for intercollegiate ice hockey. *Sport Marketing Quarterly, 16*, 106–114.

Ross, S.D., Russell, K.C., & Bang, H. (2008). An empirical assessment of spectator-based brand equity. *Journal of Sport Management, 22*, 322–337.

Ross, S.D., Walsh, P., & Maxwell, H.D. (2009). The impact of team identification on ice hockey brand associations. *International Journal of Sport Management and Marketing, 5*(1/2), 196–210.

Rossiter, J. R. (2002). The C-OAR-SE procedure for scale development in marketing. *International Journal of Research in Marketing, 19*, 305–335.

Takai, N. (2018). Shimin marason taikai gekizou no shirarezaru butaiura. [Reasons for the increase in marathon events]. Retrieved from https://toyo keizai.net/articles/-/208012

Trip Editor (2018). *Itte yokatta tabishite yokatta: Todoufukenbetsu manzokudo ranking.* [Visit and travel satisfaction in Japan] https://tripeditor.com/352687

Xing, X., & Chalip, L. (2006). Effects of hosting a sport event on destination brand: A test of co-branding and match-up models. *Sport Management Review*, *9*(1), 49–78.

Yoshida, M., Nakazawa, M., Inoue, T., Katakami, C., & Iwamura, S. (2013). [Supotsu ibento ni okeru saikansen koudou: Saikansen ito no sakihe] Repurchase behavior at sporting events: Beyond repurchase intentions. *Japanese Journal of Sport Management*, *5*(1), 3–18.

6 The socio-economic and branding impacts of an international sporting event in Japan

Le Tour de France Saitama Criterium

Daichi Oshimi

Introduction

Place event marketing is defined as "a comprehensive group of activities related to city branding and city promotion and can be part of long-term urban development and promotional strategies" (Cundy 2019, 17). In other words, place event marketing is a proactive approach in which events are used as a tool for city branding. Thus, event organisers need to leverage events to maximise their benefits on host communities (Chalip 2006). This chapter covers the Le Tour de France Saitama Criterium as a case of place event marketing. Since 2013, this one-day event has been held in commemoration of the 100th anniversary of the Tour de France with the cooperation of Saitama City. According to the event organiser, over 90,000 spectators – including local visitors and tourists – attend this event every year. Furthermore, this event has been broadcast in about 190 countries worldwide. Thus, the local government expects a socio-economic impact of the event on their city: for example through economic development, external city image enhancement, or positive word-of-mouth for their city's promotion.

In this chapter, the author gives an overview of the impacts of events on host cities presented in sports and tourism management literature and presents the results of his own studies. The main aim of this chapter is the presentation of the results of empirical research on socio-economic and image impacts of the *Le Tour de France Saitama Criterium* based on a questionnaire survey at the event site. The economic impact is calculated by using the simplified Meetings, Incentives, Conferencing, and Exhibitions (MICE) economic impact measurement model developed by the Japan Tourism Agency (JTA). The calculation is based on an Input–Output table. Social impact is investigated by utilising social impact scales from previous studies (Oshimi et al. 2016) to analyse how social impact perceptions influence behavioural intentions from the local visitor's and tourist's viewpoint. Statistical analysis (multiple-regression analysis, structural equation modelling) is conducted to clarify the social impact of sporting event on the host community.

Social, economic, and promotional impacts of sports events on host cities

Events can be classified into planned or unplanned events (Getz and Page 2016). Planned events are created for a purpose and are well publicised in advance. They are divided into several categories according to their type (cultural celebrations, political and state, arts and entertainment, business and trade, educational and scientific, sports competition, recreational and private events) and scale (local, the regional, periodic hallmark, and occasional mega-events) (Getz 2008). Saitama criterium is a planned event and can be categorised as a periodic regional event (Getz 2008).

Sporting events are one of the main products inducing sports tourism and have several tangible and intangible impacts on local cities. For example, sporting events are often assumed to have positive socio-economic and image effects (Crompton et al. 2001) including tourism-related spending (Kirkup and Major 2006; Mules 1998); urban regeneration (Gratton et al. 2005; Weed and Bull 2004); employment opportunities (Spilling 1998); psychic income (Inoue and Havard 2014; Kim and Walker 2012; Waitt 2003); image enhancement (Kim and Morrison 2005; Kim and Petrick 2005); improved quality of life (Jeong and Faulkner 1996; Kaplanidou et al. 2013); increased social capital (Heere et al. 2013; Lee et al. 2013); health literacy (Lee et al. 2013); sports participation (Collins et al. 1999); and interest in a foreign culture (Balduck et al. 2011). These impacts encourage people to support future events (Gursoy and Kendall 2006; Kaplanidou et al. 2013). However, the existence of such impacts is not always empirically supported. For example, the net economic impacts of events are often more negative when measured using a cost–benefit analysis method rather than a standard economic impact analysis (Kesenne 2005; Taks et al. 2011). There are some critics (Misener et al. 2015; Weed et al. 2009) who argue that there is no empirical evidence that sporting events stimulate new participation in sports.

In addition to the positive impacts (perceived or real), there are always the negative impacts. Examples of negative impacts include excessive spending of public funds (Deccio and Baloglu 2002; Waitt 2003); mobility problems such as traffic congestion and the difficulty experienced by residents in accessing their homes (Fredline 2005); increasing prices (Deccio and Baloglu 2002); increasing crime rates (Kim et al. 2006); prostitution and displacement of local residents (Ohmann et al. 2006); and the conflict between tourists and residents (Balduck et al. 2011). In fact, the net outcomes of sports events are uncertain, therefore, we need to obtain empirical evidence to validate the hosting sporting events. In particular, the consent of the local residents is necessary, as public funds are required to support the event (Ohmann et al. 2006), and because their perceptions of the event's impact will positively or negatively influence their intention of supporting the event (Oshimi et al. 2016).

To leverage positive impacts for host city promotion, having good relationships with various stakeholders such as event visitors and media is important (O'Brien and Chalip 2008). Media strategies are necessary to enhance a host city's image by regarding sporting events as a showcase of the host city. Furthermore, event visitors contribute to increasing event expenditures and enhancing business relationships (Chalip 2004). One unique characteristic of sporting events is the strength of fan loyalty towards their favourite team or event (Wann and Branscombe 1990). Once event organisers harness a core group of fans for their event/team, these individuals have great potential to be a compelling ambassador of their event (Funk and James 2001).

Study site – the city of Saitama

Saitama is located about 29 km north of Tokyo and has a population of approximately 1.25 million people. The city was founded in 2001 by combining three cities, and its economy is mainly constructed by various manufactures of automotive, food, optical, and pharmaceutical products. On the other hand, the city has few tourism resources and accommodation facilities. Moreover, it does have sporting facilities, such as an arena and a stadium, as well as professional sports teams including two professional soccer teams and one basketball team. For example, the soccer stadium is called Saitama Stadium 2002 (63,700 capacity) and it was constructed for the 2002 FIFA World Cup (four games were held, including a semi-final match). Another sporting facility is the multipurpose arena called Saitama Super Arena (over 19,000 capacity) and is often used for sporting events (this arena is the venue for the Tokyo 2020 basketball games), concerts, etc.

Although Saitama is among the big cities in Japan, its brand image is much worse than other big cities such as Tokyo or Yokohama (Shimizu 2015). Thus, the city mayor wanted to improve the city image by utilising their existing resource (i.e. for sports). For example, establishing the Saitama Sport Commission in 2012 is also one of the strategies for city promotion through sports.

The Saitama Sports Commission (SSC) established in the city is the first regional sports commission in Japan. Their annual budget is 683,000 USD (1 USD = 100 JPY) and the breakdown is 140,000 USD for bidding on sporting events; 150,000 USD to support sports event management; 135,000 USD for walking events; and 258,000 USD for the promotion of the Saitama Criterium in 2014. This commission has calculated the economic impact of sporting events in the city since 2011, and the result is that Saitama had hosted 157 sporting events (e.g., running events, local sporting events) that brought in 811,488 visitors (including players, spectators, staff members, etc.), which had an economic impact of 28.7 billion USD brought into the city from 2011 to 2015. In 2019, the SSC restructured their organisation from one of a city government-oriented style to a private-oriented one in order to be a more independent organisation. Thus, they need more strategy for branding and promotion of the city and its sports event.

Le Tour de France Saitama Criterium
– presentation of the event

With cooperation of the SSC, the authors began a research project to study the economic and social impacts of Le Tour de France Saitama Criterium. This one-day event is held in commemoration of the 100th anniversary of the Tour de France with the cooperation from the city of Saitama. The main organiser of this event is Saitama Sports Commission, and the co-organisers are Saitama Prefecture, Saitama City, Saitama Tourism and International Relations Bureau, and Amaury Sport Organisation (A.S.O.). The event has been held for six consecutive years since 2013 and it has a contract to hold the event until 2021. The race is set up on the closed-circuit in the centre of Saitama City by using existing facilities (e.g. public roads). Top international cyclists, who have won various categories of Tour de France, and leading Japanese cyclists participate in the event. Several entertainment programmes about French culture were held nearby during the event. Moreover, French food, wine, and several specialty products were marked with the event's logo and sold. This helped create an international atmosphere that ensured that the spectators enjoyed themselves. This event was the first cycling competition named after the Tour de France. The brand Tour de France is franchised for expanding markets through the world by the French organiser. Furthermore, using the brand is a perfect tool for promoting the city of Saitama, by utilising the famous name Tour de France.

According to the event organiser, over 90,000 spectators attend the event every year. For further development of Tour de France Saitama Criterium, orgnanisers set up a new sales promotion strategy such as hospitality tickets for VIPs and executives. One ticket costs 5,000 USD and includes such attractions as meeting with the players, attending an exclusive VIP party, access to a VIP seat, accommodation with dinner and breakfast, etc. This event already has a good public reputation and it could reach the next level by adding high-end (premier) images (e.g. elegant, gorgeous) to the event. Furthermore, leveraging this event for city projects such as infrastructure development for local cyclists and people outside the city are a good way for place branding such as "sport city", "cycle city", or "healthy city".

Study methods

The data for the socio-economic impact study were collected at the 2015 Le Tour de France Saitama Criterium. A well-trained staff of 15 individuals conducted a questionnaire survey among the spectators at the event site. Convenience sampling was used, and 860 usable questionnaires were obtained. Fifty-seven per cent of respondents were male and the average age of the respondents was 38.2. Thirty per cent of respondents were from Saitama city and others from outside the city. According to the event organiser, the number of spectators in attendance was 95,000 this year. The economic impact was calculated by using the simplified MICE economic impact measurement model developed by the JTA. The calculation is based on an Input–Output table. Table 6.1. shows the spectators' average spending at the event. The sample was divided into two groups with one consisting of spectators who lived in the host city/neighbouring areas and the other consisting of those who lived outside the host city. This is because the number of expenses for accommodation, travel, etc. is usually determined by the distance from home to a destination.

Table 6.1 Spectators' average spending at the event (USD)

Residence	Accommodation	Food	Shopping/ souvenirs	Sightseeing/ entertainment	Others	Total expense	Ratio (%)	n
Host city (including neighbouring areas)	0.5	26.2	19.5	5.6	2.2	54.0	87.5	83,125
Outside host city	6.5	31.7	47.6	8.4	8.0	102.2	12.5	11,875
Total							100.0	95,000

Source: Author's elaboration.

Regarding social impact analysis, two types of research were conducted. First, multiple regression analysis was conducted to examine how spectators' (i.e. local visitors and tourists) perceived social impacts of the event influence their attitude towards the host city. The social impact scale (6 factors and 21 items) was utilised, which includes "Economic Development", "Tourism Development", "Cultural Interest and New Opportunity", "External Image Enhancement", "Consolidation and Pride", and "Disorder and Congestion" based on previous research (Oshimi et al. 2016). Most social impact studies analyse the perceptions of local visitors or tourists towards sporting events based on a variety of impacts and address their perceptions, such as economic/tourism development, cultural interest, and image enhancement (Balduck et al. 2011; Kim and Petrick 2005; Ma et al. 2013).

Further on, Structural Equation Modeling (SEM) and multiple regression analysis were conducted to analyse the influence of host city image on behavioural intentions (e.g. revisiting and word-of-mouth), including the other related factors. City image is one of the social impacts and is utilised in the initial phase of brand development, which requires two perspectives: the tourists' and the residents' (Kavaratzis 2004). Several empirical studies clarified the positive relationships between host city image and behavioural intentions from its visitors' (e.g. Chen and Funk 2010; Kaplanidou et al. 2012; Papadimitriou et al. 2015) and residents' perspective (e.g. Oshimi and Harada 2019). Recent research suggests that residents' well-being has been a primary concern, especially for policymakers who strive to utilise the power of tourism (Yolal et al. 2016). Furthermore, event organisers could expect residents to be destination ambassadors (Hudson and Hawkins 2006; Schroeder 1996). Thus, analysing event impact from both perspectives (i.e. local visitors and tourists) is necessary for place event marketing. We adopted a "City Image" scale (21 items; Oshimi and Harada 2019) with the following factors: "City/ Convenience" – four items; "Sports" – four items; "City Atmosphere" – four items; "Sightseeing/Leisure" – three items; "Nature" – three items; "Business" – three items; "Event Excitement" – three items (Lacey and Close 2013); and "Behavioural Intentions" (two items): "Would you recommend the event to others (e.g. friends, family)?" and "Would you want to revisit this event next year?" A seven-point Likert scale ranging from one (absolutely disagree) to seven (absolutely agree) was adapted in all scales in social impact analysis.

Research results

Economic impact

The result of the economic impact of the event is presented in Table 6.2. The total economic ripple effect was 2.51 billion USD. Furthermore, additional tax income was generated at the national level (13.7 million USD), the province level (9.7 million USD), and the municipal level (6.6 million USD). Moreover, there was an employment inducement of 47,579 (persons/day).

Table 6.2 The economic ripple effect of *Le Tour de France Saitama Criterium* 2016

Index	Total amount	Unit
Total Expenditure	12.1	million USD
1. Direct impacts	11.5	million USD
2. Indirect impacts	8.3	million USD
3. Induced impacts	5.2	million USD
Economic ripple effect (1+2+3)	**25.1**	million USD
National tax	13.7	million USD
Province tax	9.7	million USD
Municipal tax	6.6	million USD
Employment inducement	47,579	person/day

Source: Author's elaboration.

Social impact

Multiple regression analysis of the perceived social impacts on positive attitudes toward the host city showed that Economic Development and Cultural Interest and New Opportunity had a positive impact on the spectators' attitudes towards the host city (See Table 6.3.). These results are in line with the results of studies regarding previous events (Oshimi et al. 2016). These factors are useful variables for determining whether spectators have positive attitudes towards host cities/ events. For example, as mentioned in the previous section, Le Tour de France Saitama Criterium has held several entertainment programmes related to the culture of France during the event near its site. This enabled spectators to have a cultural experience during the event, which helped them develop positive attitudes towards the host city. However, considering the negative influence of Disorder and Congestion, event managers should focus on perceptions of Economic Development and Cultural Interest and New Opportunity, while simultaneously minimising the impacts of Disorder and Congestion.

Table 6.3 The results of multiple regression analysis of the perceived social impacts on positive attitudes towards the host city

Factors	B	P
Economic Development	**0.235**	0.005
Tourism Development	-0.089	0.333
Cultural Interest and New Opportunity	**0.256**	0.001
External Image Enhancement	0.078	0.313
Consolidation and Pride	0.096	0.300
Disorder and Congestion	**-0.151**	0.022
R^2	0.267	

Source: Author's elaboration.

Table 6.4 The results of multiple regression analysis of host city image on behavioural intention between local visitors and tourists

Factors	Local visitor		Tourist	
	β	P	β	P
City/Convenience	0.02	0.81	0.22	0.03
Sports	0.31	0.00	0.18	0.03
City atmosphere	0.06	0.52	0.00	0.97
Sightseeing/Leisure	0.16	0.03	0.01	0.91
Nature	−0.03	0.62	0.20	0.01
Business	0.02	0.84	−0.06	0.50
R²	0.20		0.21	

Source: Author's elaboration.

Social impact: city image perspective

Table 6.4 shows the influence of city images on behavioural intentions. The results were separated between local visitors and tourists. A common result in both segments was "Sport" had a positive influence on behavioural intentions. It indicates that sport image could be a useful tool for event/city promotion towards event spectators by fostering their loyalty behaviour (positive word-of-mouth and re-visit intention). An interesting result is the heterogeneous results between two segments (i.e. local visitors and tourists), which is City/Convenience and Nature image are influential for local visitors, while Sightseeing/Leisure image is for tourists. For local visitors, the factors related to their life (i.e. Convenience and Nature) have a more powerful contribution to behavioural intentions because event site is their hometown. On the other hand, it does make sense for tourists to evaluate more Sightseeing/Leisure than other factors because they expect enjoyment toward their destination.

The relationship between "Event Excitement", "City Image", and "Behavioural Intention" is presented in Figure 6.1. Event excitement was added to the model by considering the emotional aspects of a sporting event (e.g., Uhrich and Benkenstein 2012). As a result, event excitement more positively influenced the intention to visit the city than did the city image. Furthermore, event excitement positively influenced the city image. This indicates that positive emotion is an important factor to consider for behavioural intention as well as for city image improvement (i.e. city branding).

Conclusions

Summarising, this case is prominent because the city of Saitama made an effort to show the impact of sporting events to validate holding such events. Sporting

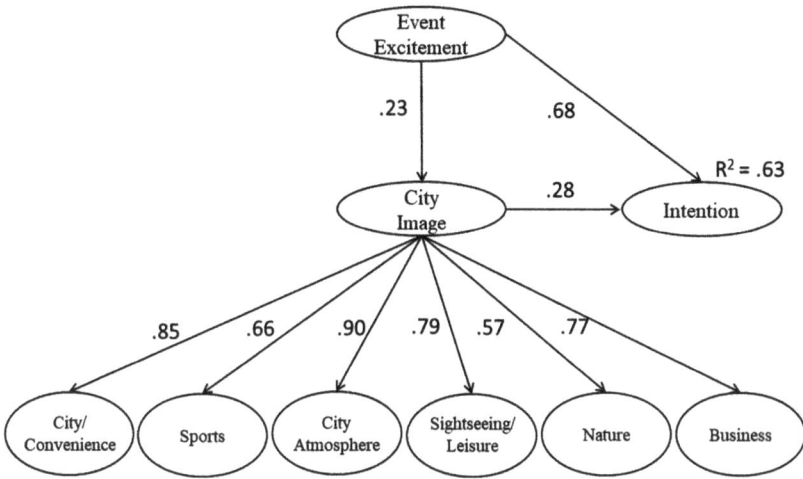

Figure 6.1 Structural equation modelling results. Notes: (χ2(df) 969.19 (315), $p < .001$, CFI =.923, TLI = .914, RMSEA =.067, SRMR = .055). Source: Author elaboration.

events should be hosted with the consent of several stakeholders (e.g. local residents) because public funds are required to support these events. Thus, local governments, event organisers, or both have to show that such events have benefits, not only in terms of economic impact but also in terms of social impact including place branding and place marketing. Empirical results suggest that perceived social impacts contribute to the positive attitude towards the host city and the event itself. Especially, perceived "economic development", "cultural interest and new opportunity", "city image", and "event excitement" could be influential tools for city promotion as well as for positive behavioural intention towards the host city and event. Since these influences could be changed by subjects (i.e. local visitors and tourists), event organisers need to make an effort to consider optimising their existing resources to maximise the potential benefits of the event by analysing marketing data. This is one of the place event marketing activities and a necessary process to leverage sporting events for the host community (Chalip 2006).

References

Balduck, A.L., Maes, M., & Buelens, M. (2011). The social impact of the Tour de France: Comparisons of residents' pre-and post-event perceptions. *European Sport Management Quarterly, 11*, 91–113. Doi:10.1080/16184742.2011.559134

Chalip, L. (2004). Beyond impact: A general model for host community event leverage. In: Ritchie, B.W. and Adair, D. (eds) *Sport Tourism: Interrelationships, Impacts and Issues*. Clevedon, UK: Channel View.

Chalip, L. (2006). Towards social leverage of sport events. *Journal of Sport and Tourism*, *11*, 109–127. Doi:10.1080/14775080601155126

Chen, N., & Funk, D.C. (2010). Exploring destination image, experience and revisit intention: A comparison of sport and non-sport tourist perceptions. *Journal of Sport & Tourism*, *15*(3), 239–259. Doi:10.1080/14775085.2010.513148

Collins, M.F., Henry, I.P., Houlihan, B., & Buller, J. (1999). *Research Report: Sport and Social Exclusion*. Loughborough, UK: Institute of Sport and Leisure Policy, Loughborough University.

Crompton, J.L., Lee, S., & Shuster, T.S. (2001). A guide for undertaking economic impact studies: The Springfest example. *Journal of Travel Research*, *40*, 79–87. Doi:10.1177/004728750104000110

Cudny, W. (2019). *Urban Events, Place Branding and Promotion: Place Event Marketing*. London: Routledge.

Deccio, C., & Baloglu, S. (2002). Nonhost community resident reactions to the 2002 Winter Olympics: The spillover impacts. *Journal of Travel Research*, *41*(1), 46–56. Doi:10.1177/0047287502041001006

Fredline, E. (2005). Host and guest relations and sport tourism. *Sport in Society*, *8*, 263–279. Doi:10.1080/17430430500087328

Funk, D.C., & James, J.D. (2001). The psychological continuum model: A conceptual framework for understanding an individual's psychological connection to sport. *Sport Management Review*, *4*, 119–150.

Getz, D. (2008). Event tourism: Definition, evolution, and research. *Tourism Management*, *29*(3), 403–428. Doi:10.1016/j.tourman.2007.07.017

Getz, D., & Page, S.J. (2016). *Event Studies: Theory, Research and Policy for Planned Events*. London and New York: Routledge.

Gratton, C., Shibli, S., & Coleman, R. (2005). The economics of sport tourism at major sports events. In J. Higham (Ed.), *Sport Tourism Destinations: Issues, Opportunities, and Analysis* (pp. 233–247). Oxford, UK: Elsevier.

Gursoy, D., & Kendall, K.W. (2006). Hosting mega events: Modeling locals' support. *Annals of Tourism Research*, *33*, 603–623. Doi:10.1016/j.annals.2006.01.005

Heere, B., Walker, M., Gibson, H., Thapa, B., Geldenhuys, S., & Coetzee, W. (2013). The power to unite a nation: The social value of the 2010 FIFA World Cup in South Africa. *European Sport Management Quarterly*, *13*, 450–471. Doi:10.1080/16184742.2013.809136

Hudson, M., & Hawkins, N. (2006). A tale of two cities: A commentary on historic and current marketing strategies used by the Liverpool and Glasgow regions. *Place Branding*, *2*(2), 155–176.

Inoue, Y., & Havard, C.T. (2014). Determinants and consequences of the perceived social impact of a sport event. *Journal of Sport Management*, *28*, 295–310. Doi:10.1123/jsm.2013-0136

Jeong, G., & Faulkner, B. (1996). Resident perceptions of mega-event impacts: The Taejon International exposition case. *Festival Management and Event Tourism*, *4*(1), 3–11. Doi:10.3727/106527096792232388

Kaplanidou, K., Jordan, J.S., Funk, D.C., & Rindinger, L.L. (2012). Recurring sport events and destination image perceptions: Impact on active sport tourist behavioral intentions and place attachment. *Journal of Sport Management*, *26*, 237–248.

Kaplanidou, K., Karadakis, K., Gibson, H., Thapa, B., Walker, M., Geldenhuys, S., & Coetzee, W. (2013). Quality of life, event impacts, and mega-event support among South African residents before and after the 2010 FIFA World Cup. *Journal of Travel Research*, *52*, 631–645. Doi:10.1177/0047287513478501

Kavaratzis, M. (2004). From city marketing to city branding: Towards a theoretical framework for developing city brands. *Place Branding, 1*(1), 58–73.

Kesenne, S. (2005). Do we need an economic impact study or a cost-benefit analysis of a sports event? *European Sport Management Quarterly, 5*, 133–142.

Kim, H.J., Gursoy, D., & Lee, S.-B. (2006). The impact of the 2002 World Cup on South Korea: Comparisons of pre- and post-games. *Tourism Management, 27*, 86–96. Doi:10.1016/j.tourman.2004.07.010

Kim, S.S., & Morrison, A.M. (2005). Change of images of South Korea among foreign tourists after the 2002 FIFA World Cup. *Tourism Management, 26*, 233–47. Doi:10.1016/j.tourman.2003.11.003

Kim, S.S., & Petrick, J.F. (2005). Residents' perceptions on impacts of the FIFA 2002 World Cup: The case of Seoul as a host city. *Tourism Management, 26*, 25–38. Doi:10.1016/j.tourman.2003.09.013

Kim, W., & Walker, M. (2012). Measuring the social impacts associated with Super Bowl XLIII: Preliminary development of a psychic income scale. *Sport Management Review, 15*(1), 91–108. Doi:10.1016/j.smr.2011.05.007

Kirkup, N., & Major, B. (2006). The reliability of economic impact studies of the Olympic Games: A post games study of Sydney 2000 and considerations for London 2012. *Journal of Sport and Tourism, 11*, 275–296. Doi:10.1080/14775080701400943

Lacey, R., & Close, A.G. (2013). How fit connects service brand sponsors with consumers' passions for sponsored events. *International Journal of Sports Marketing & Sponsorship, 14*(3): 212–228. Doi:10.1108/IJSMS-14-03-2013-B005

Lee, S., Cornwell, T., & Babiak, K. (2013). Developing an instrument to measure the social impact of sport: Social capital, collective identities, health literacy, well-being, and human capital. *Journal of Sport Management, 27*(1), 24–42.

Ma, S.C., Ma, S.M., Wu, J.H., & Rotherham, I.D. (2013). Host residents' perception changes on major sport events. *European Sport Management Quarterly, 13*, 511–536. Doi:10.1080/16184742.2013.838980

Misener, L., Taks, M., Chalip, L., & Green, C. (2015). The elusive "trickle down effect" of sport events: Assumptions and missed opportunities. *Managing Leisure, 20*(2), 135–156.

Mules, T. (1998). Events tourism and economic development in Australia. In D. Tyler, Y.Guerrier, & M. Robertson (Eds.), *Managing Tourism in Cities: Policy, Process, and Practice*. Chichester, UK: Wiley.

O'brien, D., & Chalip, L. (2008). Sport events and strategic leveraging: Pushing towards the triple bottom line. In: Woodside, A.G., and Martin, D. (eds) *Tourism Management: Analysis, Behaviour, and Strategy*. Wallingford, Oxford: CAB eBooks.

Ohmann, S., Jones, I., & Wilkes, K. (2006). The perceived social impacts of the 2006 Football World Cup on Munich residents. *Journal of Sport and Tourism, 11*, 129–152. Doi:10.1080/14775080601155167

Oshimi, D., & Harada, M (2019). Host resident's role in sporting events: The city image perspective. *Sport Management Review, 22*(2), 263–275. Doi:10.1016/j.smr.2018.04.002

Oshimi, D., Harada, M., & Fukuhara, T. (2016). Residents' perceptions on the socio-economic impacts of an international sporting event: Applying panel data design and a moderate variable. *Journal of Convention & Event Tourism, 17*(4), 294–317. Doi:10.1080/15470148.2016.1142919

Papadimitriou, D., Kaplanidou, K., & Apostolopoulou, A. (2015). Destination image components and word-of-mouth intentions in urban tourism: A multigroup approach. *Journal of Hospitality & Tourism Research*, 5(May), 1–21. Doi:10.1177/1096348015584443

Schroeder, T. (1996). The relationship of residents' image of their state as a tourist destination and their support for tourism. *Journal of Travel Resources*, *34*(4), 71–73.

Shimizu, H. (2015). *Sport City Saitama*. Saitama, Japan: Saitama Newspaper.

Spilling, O.R. (1998). Beyond intermezzo? On the long-term industrial impacts of mega-events: The case of Lillehammer 1994. *Festival Management & Event Tourism*, *5*, 101–122. Retrieved from https://www.cognizantcommunication.com/journals-previou slypublished/festival-management-a-event-tourism

Taks, M., Kesenne, S., Chalip, L., Green, B.C., & Martyn, S. (2011). Economic impact study versus cost-benefit analysis: An empirical example of a medium sized international sporting event. *International Journal of Sport Finances*, *6*, 187–203.

Uhrich, S., & Benkenstein, M. (2012). Physical and social atmospheric effects in hedonic service consumption: Customers' roles at sporting events. *Service Industries Journal*, *32*(11): 1741–1757. Doi: 10.1080/02642069.2011.556190

Waitt, G. (2003). Social impacts of the Sydney Olympics. *Annals of Tourism Research*, *30*(1), 194–215. Doi:10.1016/S0160-7383(02)00050-6

Wann, D.L., & Branscombe, N.R. (1990). Die-hard and fair-weather fans: Effects of identification on BIRGing and CORFing tendencies. *Journal of Sport and Social Issues*, *14*, 103–117.

Weed, M., & Bull, C. (2004). *Sports Tourism: Participants, Policy, and Providers*. Oxford, UK: Elsevier Butterworth Heinemann.

Weed, M., Coren, E., & Fiore, J. (2009). *A Systematic Review of the Evidence Base for Developing a Physical Activity and Health Legacy from the London 2012 Olympic and Paralympic Games*. Canterbury, UK: Canterbury Christ Church University.

Yolal, M., Gursoy, D., Uysal, M., Kim, H.L., & Karacaoğlu, S. (2016). Impacts of festivals and events on residents' well-being. *Annals of Tourism Research*, *61*, 1–18.

7 From religious festival to cultural carnival

Durga Puja and city branding of Kolkata, India

Aparajita De

Introduction

> I was born in Uttar Pradesh. Brought up in Uttar Pradesh and then I went to a college in…well not in Uttar Pradesh. It was a city in Eastern India and by the time my first semester was half done the college was shut down for ten days of vacation. It was strange because Diwali had not arrived yet and Holi was of course too far away. Back in North when we were all fasting on Navratri and burning Ravan effigy on Dussehera the people in the East was celebrating something else and that is when I came to know about Durga Puja…In this world I saw thousands of people hopping the pandals through the night something I would not have imagined otherwise. The lights of these pandals take creativity beyond the puja beyond the myths. I listen to the music playing all over the city which took me into the world of stories. And now I can say that if one wants to understand India one should start from Kolkata because where else can we get to see an India that we have read in the books.
>
> (Ethereal: My First Kolkata Durga Puja)

The above narrative is from a YouTube video, My First Kolkata Durga Puja,[1] that was widely circulated on social media showcasing a montage of common scenes during the festivities in Kolkata when the city experiences a carnivalesque transformation dotted with innumerable but massive *pandals* or marquee (i.e. temporary structures made from bamboo and covered with cloth) that may be built reflecting a particular theme, with equally impressive idols of Goddess Durga, light decorations, stalls selling different kinds of things from food to handicrafts, the sound of *aarti* (ritualistic worship of Goddess Durga), *dhaak* (drums) and *kashor ghonta* (bronze bell and gong), crowded streets with people dressed in their best hopping from one puja pandal to the other. Durga Puja, is a six-day-long annual Hindu festival that reveres the Goddess Durga; having originated in the colonial period, Durga Puja has over the years emerged as one of the most popular and major festivals that the city celebrates. The video itself had nearly two hundred thousand views and with a number of reaction videos[2] covered by YouTubers across the world. The reactions mostly comment on the artistry, creativity, and vibrancy involved in the festivities rather than the rituals and the religious practice of Durga Puja. Most of the YouTube commentators draw attention to the ability of the video to reproduce the feel and experience of the Durga Puja making them want to visit Kolkata during the event. In fact, the comment section of the reaction videos has innumerable invites posted to the YouTubers to physically visit and experience Kolkata's Durga Puja.

It is interesting to note that one of the YouTube commentators on Ethereal video remarks that such social media coverage not only opened up his thinking on India which was very different from how conventional media represents India, emphasising only particular narratives.[3] Yet when it comes to an event like Durga Puja in Kolkata the coverage of it by the national media[4] is no different and highlights "the burst of creativity" reflected in the themes of the pandals of Durga puja ranging from fake news in social media, Balakot airstrike[5] to the politically controversial National Registry of Citizenship.[6] The story of the letter and how it has evolved with technology. Suraj Mohan Jha from Lok Sabha TV in his documentary on Durga Puja[7] stresses that what distinguishes the Durga Puja in Kolkata from other festivities is its cultural creativity and the discovery of the City of Kolkata through the Durga Puja festivities. This brings the key research issues that this chapter focuses on.

1. How Durga Puja, an essentially religious festival, can create an image and brand the city of Kolkata, as a site of cultural creativity.
2. What is the role of social media in this branding process.

The chapter adopts a multi-method approach relying primarily on qualitative research methods that includes literature analysis, content analysis of media coverage on Durga Puja, direct observation of the event, in-depth interviews, and digital ethnography of Facebook, blogs, YouTube channels, and Instagram were conducted. Most of the fieldwork and digital ethnography was done from 2017 to 2019. The digital ethnography included the analysis of Facebook pages and posts on Durga Puja that were public and non-moderated, the reason being that all posts were open to the public ensuring that anyone could access these pages and posts. Also, Facebook pages that were non-moderated were chosen so that all kinds of comments can be seen by everyone and that negative comments are not moderated or unpublished. Similarly, hashtags on Durga Puja, Kolkata, Puja parikrama (pandal hopping) were followed on Instagram, and accounts on Durga Puja and Kolkata with over 5,000 followers were considered.

Since Durga Puja is held in over 3,000 sites in the city of Kolkata it is not possible to carry fieldwork at every site. Therefore about 10 specific sites were chosen for the purpose of fieldwork. The sites included the space of the pandals and adjoining areas and streets. These sites were selected based on their popularity; the scale of the puja in terms of expenditures made towards organising the Durga Puja; and the number of years for which the puja has been organised. A total of 25 in-depth interviews were taken that include Durga Puja organisers, tour operators, idol makers, decorators involved in the making of pandals, street vendors, local residents, non-resident Bengalis, and tourists.

A multi-sited digital ethnography was undertaken that covered Facebook pages and posts, Instagram hashtags and accounts, blogs, vlogs, and videos uploaded on YouTube on Durga Puja. The core questions of the field focused on how people perceived Durga Puja as a festival and how this was reproduced and communicated as part of Kolkata's identity and the opinions people formed about the event and Kolkata as a whole. The fieldwork was conducted to capture how the branding process of Kolkata took place in social media and how Kolkata's image and the brand have evolved and are being circulated.

One of the interviewees Amit, an advertising professional and a Bengali who grew up in Kolkata and a resident of Mumbai for the past two decades, says:

> In today's time you no longer have to live in your own city to be reminded of the festivities. Take for example Durga Puja one can hardly forget it you have the advertisements and social media, posts on Facebook, YouTube, Instagram, the memes and the messages in WhatsApp that bombard you with it and tell you that Puja is around the corner rather than the Bengali calendar or the changing of the seasons – the powdery blue skies and *kash phool* (catkins or Kans grass) that is normally associated with the Puja.

Amit's narrative brings to light how social media has become an integral part of our lives connecting us to events and places.

Social media as Hoskins (2017) argues is not only omnipresent in our everyday lives but makes possible a connective turn that connects us to people, places, and events. Navpreet Arora, owner of fun on streets a walking tour company underlining this connective turn made possible by social media informs that on reading her blogs and her posts on various Facebook pages a group of American women had particularly contacted her for one of her walking tours of the "Bonedi bari puja", the traditional Durga Pujas that are held at the homes of the rich elites of Northern Kolkata dating back to the colonial era. She adds that they even wore the traditional red and white saree worn by Bengali women for the pujas.[8] Navpreet believes that "social media has generated great buzz around Durga puja in Kolkata and has created a certain brand value for Kolkata which already has and can have a far-reaching economic implication for Kolkata and the State". Thus, it is of little surprise to see almost all top corporate companies not only sponsoring Durga Puja in Kolkata but having special advertisement campaigns from FMCG companies like ITC, Pepsi, to media houses like Times of India, financial service providers like Edelweiss and State Bank of India, retail industry leaders like Pantaloons, Myntra, Shoppers Stop, to paint companies like Asian Paints, and Kansai Nerolac.

Mukherjee (2019)[9] notes that "none of these [advertisements] pseudo-aesthetic videos seem to be advertising any product or products. Instead, they advertise Durga Puja". In a way, it is indicative of how Durga Puja slowly yet steadfastly has emerged as an "event" that can not only be utilised in the branding of products and creating brand images of corporate companies but that of the city of Kolkata as well. The State government having realised the potential of Durga Puja in 2016 started organising *Bhishorjon* carnival that showcased the heritage and cultural aspects of the festival indicating the active role of the State government in branding Kolkata through the festival of Durga Puja. In 2018 the Department of Tourism of the West Bengal government in collaboration with the British Council held a photography exhibition of Durga Puja in London's Totally Thames Festival to demonstrate the "experiential tourism" that Durga Puja had to offer to international tourists. The official statement of the exhibition to highlight this unique experience states that "We see a Durga Puja transcended from its religious origins to offer a broad canvas that binds communities together through artistic expression, incorporating music, theatre, fashion, and food".[10] Clearly, the State government of West Bengal thought of Durga Puja as having the potential to create economic opportunities for the city of Kolkata and the state and since 2018 has announced a total grant of 28 crores to Durga puja committees for community development programmes; waiving heavy license fees levied by the Kolkata Municipal Corporation and discounts for power consumption by the puja pandals.

The spectacles of Durga Puja and Kolkata

Grand and spectacular celebrations of Durga Puja and its association with the city of Kolkata, the capital of the eastern state of West Bengal in India, can be traced back to the eighteenth century and the emergence of the British colonial empire in South Asia. Durga Puja was celebrated initially at the homes of the rich and elite and along with ritualistic worship of the goddess Durga which came to be known for elaborate feasting and different kinds of cultural performances from dance and singing performances or nautches, mimicry or swang, puppetry, folk theatre or Jatra (Ghosh, 2006). Most of the earliest Durga Pujas were celebrated by the rich Bengali merchant class and administrators in the colonial bureaucracy to entertain the distinguished British colonisers, where each is said to have tried to outclass the other in terms of opulent displays and the splendor with which it was celebrated (Sircar, 2011).

In fact, the most notable and perhaps the oldest Durga Puja was celebrated by Raja Nabakrishna Deb of Shovabazar in veneration of the British winning over the Nawab Siraj-ud-Daulah at the Battle of Plassey around 1757 (Banerjee, 2004). Ordinary residents of Kolkata did have limited access to the Durga Pujas but it was solely restricted to the households of the elite. But as the popularity of the event grew it became more of public worship of the Durga where the entire community got together to celebrate it. According to rough estimates, the city in 2019 had 3,000, and some even claim the figure to be 3,700 puja pandals or marquee, i.e. temporary structures, that were constructed in public spaces to celebrate Durga Puja. Raju, one of the respondents interviewed for the purposes of this chapter, who organises a typical *para* (locality) puja in Kolkata describing how Durga Puja has evolved recalls:

> Earlier in Kolkata till the early 80s we used to have something like a baroari puja, where all the residents of the para (locality) used to contribute asper their means and a puja used to be organized. Our compound puja also had a

humble beginning where all the residents contributed and we held the Durga Puja in one of the empty garages. The residents themselves used to volunteer and form various committees, like the finance and pratima committee, puja committee, and cultural committee, which oversaw the entire organizing and functioning of the Puja. In the 80s some of the larger pujas were organized by the local clubs that had connections, particularly political ones, and were able to bring in sponsors and get permissions to hold pujas on a large scale. Today, the pujas are not commercialized but are corporatized like the 5 Crore sponsorship of the Deshapriya Park Puja by the Star cement company, with an 80-foot tall idol made of cement and the tallest idol of Durga puja ever built in the world!

Durga Pujas is an autumnal festival that is observed in the month of Ashwin according to the Hindu calendar, coinciding roughly with the months of September and October in the Gregorian calendar (Bandhyapadhay, 2017). The annual festival is usually celebrated with the making of pandals within which elaborately decorated ensemble of unfired clay models of the Goddess riding on a lion and with ten hands each carrying weapons especially gifted by the gods to destroy Mahisasur, the buffalo demon, who is always at her foot; and accompanying the Goddess is also her four children – goddesses Laxmi and Sarawati, and gods Ganesha and Kartick (Bean, 2012). The ten-day annual festival is marked by two simultaneous narratives – one, the long-drawn battle between the Maa Durga and Mahisasur, in which the Goddess ultimately slays the demon solemnising the victory of good over evil (Ray, 2017). The second narrative coincides with the harvesting season in Bengal and with the annual visit of the Goddess along with her children to her parent's home, and celebrations are carried out to ensure the protection and well-being of the devotees (Chakrabarti, 2001).

Ray (2017) claims that the city of Kolkata during the pujas transforms into a "spectatorial complex" that exhibits pandals and idols taking many forms from grand palaces to temples, to run-down mansions to spaceships, churches, castles, the Titanic, and Hogwarts. Ray also argues that it is this transformation into a magical fantasy land that gives the Durga puja in Kolkata a cosmopolitan character. At the same time, it added to the carnivalesque flavor with hordes of people walking or driving from one pandal to the other till late in the night and often till early mornings, enjoying the food.

Bhattacharya (2007) on the other hand argues that the ceremonies of Durga Puja in the late 19th century and early 20th century reflected a secularisation of the festival and shaping its modern form that is closely associated with the city of Kolkata. The grand scale of the pujas during that time along with the performances of popular dance, music, folk theatre created new urban sensibilities and urbanity that included and ensured the participation of people cutting across different social strata and irrespective of caste, religion, and gender. Bhattacharya (2007, p. 948) also points out that the popular songs, dances performed by nautch girls, and jatras or folk theatre were often bawdy, with underlining sexual references that were hardly reverential to Durga Puja as a religious practice or festival and the "entertainments themselves were crucial in denoting a secular urban status to religion".

Ghosh (2006) notes that historically Durga Puja was associated in equal measure with cultural practices as with articulation of public opinion on contemporary issues through the artistic rendition of the puja pandals, its decoration, and lighting. He gives examples of two particular Durga Pujas organised Sanatan Dharmotshahini Sabha and Simla Byam Samity (a gymnasium club which fostered a physical culture among the nationalist youth) which were associated with nationalist movement and expressed their nationalist leanings through covert decorating motifs and the display of martial arts prowess on *ashtami* (the eighth day of the Durga Puja). Connecting Durga Puja with contemporary issues is a trend that to date continues but it also symbolises the urban, modern, and secular character of the Durga Puja in Kolkata.

Sonali, a resident of Kolkata, stated during an interview that,

> Over the years Durga Puja and the artistry and creativity that one can witness in the making of pandals, the idols of the Goddesses, lighting and the choice of themes have been phenomenal. With each passing year, it has been grander, bigger than life... the themes have been more sensitive and really touches the concerns of the everyday man from issues concerning the environment to political controversies, to social concerns like woman's empowerment. It was a proud moment when the Durga Puja pandal of FD block – a replica of the Hogwarts school from Harry Potter drew the attention of Rowling,[11] who filed a case of copyright infringement against the Puja committee. I call it a proud moment because in a way it is a recognition for the expertise and craftsmanship of our artisans, from the idol makers to the pandal makers, electricians, *shola* makers, the *dhakis* (drum players) ...the whole army of people who makes possible the celebration of Durga Puja in such a huge manner. I think Durga Puja has not only become bigger and better but its innovativeness that stands out.... which is primarily cultural in nature and has come to be closely associated with Kolkata only.

It leaves no doubt that the Durga Puja of Kolkata remains as spectacular as it had when it was initiated and gained a mass following in the colonial era. But also what has been central to celebrations of Durga Puja in Kolkata has been certain understanding and practices of modern values whether it was Nabakrishna Deb opening the doors of his home and inviting Jews, Muslims, other Hindus, and the British on the occasion of a Hindu religious festival nor restricting its celebration to rituals alone but giving it a cultural turn through elaborate feasting, dancing, and singing and have grown into a cultural carnival.

Festivalisation of Durga Puja and the urban identity of Kolkata

Tell outsiders about the importance of Puja in Calcutta and they'll scoff. Don't be silly, they'll say. Puja is a religious festival. And Bengal has voted for the CPM

(*The Communist Party of India*) since 1977. How can godless Bengal be so hung up on a religious festival? I never know how to explain them that to a Bengali, religion consists of much more than shouting Jai Shri Ram or pulling down somebody's mosque. It has little to do with meaningless ritual or sinister political activity. The essence of Puja is that all the passions of Bengal converge: emotion, culture, the love of life, the warmth of being together, the joy of celebration, the pride in artistic expression and yes, the cult of the goddess. It may be about religion. But is about much more than just worship.[12]

Vir Sanghvi openly admits on his page that he wrote two pieces on Durga Puja in the newspaper *Hindustan Times* more than a decade back which he had forgotten till it went viral on social media several years back.[13] Since then each year versions of his original article circulates on social media which deeply resonates with the emotional experience of Durga Puja in Kolkata that most Bengalis share. Many posted and commented on how the article brought out the *Bangaliana* (Bengaliness) or the fact that "no big carnival in any corner of the world can match up to it" and "one's heart beats for her [Kolkata] and practically sprints during Pujo".[14] The Durga Puja in Kolkata as a festival creates an almost magical experience and that which cannot be experienced elsewhere as Vir Sanghvi writes that it may be about religion but goes much beyond it to encompass culture, artistic expression, emotion, warmth of being together, the celebration of joy, and love for life.

Durga Puja brings the Bengali community together based on shared values, their sense of identity and pride as Bengalis and in the city of Kolkata. Durga Puja as a festival is a unique leisure and cultural experience that can motivate people to come and participate and at the same time strengthen the feeling of community belonging, pride, and a sense of local identity as other festivals do (see: Getz, 1991; Bennett and Woodward, 2014). In other words, Durga Puja can create the essence of Kolkata-Bengali culture on one hand, and on the other it has been able to produce spaces for encountering, consuming, and experiencing this unique and distinctive culture. An argument that Bennett points out is that festivals become the popular means through which local cultures are consumed and experienced in the larger global contexts (Bennett et al., 2014).

The relevance of the Durga Puja as a major event can be understood when the Home Minister of India and the Chief Minister of West Bengal inaugurates the Durga Puja in Kolkata. Nor is it surprising when newspaper reports claim that,

> estimating the size of the Durga Puja economy seems like a debatable exercise, a 2013 Assocham report titled "West Bengal cashing in on Durga Puja celebrations" pegged it at Rs 25,000 crore and growing at about 35 per cent CAGR. It projected its size to be Rs 40,000 crore by 2015. If that figure were to be extrapolated to 2018, the size would be Rs 1.12 lakh crore and Rs 1.5 lakh crore in 2019. As West Bengal's current GDP is Rs 10.20 trillion, as per the estimate, the Puja economy contributes a little over 10 per cent to the state's GDP.[15]

It is not far from the truth when social media posts claim that Durga Puja in Kolkata has fast progressed into a global cultural event today having a great economic impact but has in fact been unaffected by the economic recession.[16] In fact, one of the social media posts quoting the survey conducted by Associated Chambers of Commerce and Industry of India (ASSOCHAM), states that Durga Puja market has been growing at a compound annual growth rate of about 35 per cent. Explaining the economy of the Puja, Gayen in his Facebook post adds that instead of being an "enormous wastage … this money flows back to the economy like any other spends, thus generates more spends by those who actually earn it. In decaying economy of West Bengal such spends are kinds of blessings".[17] At the same time, Chandra Pal, a third-generation idol maker from Kumortoli, informs that

> Durga Puja in Kolkata created a means of employment for a whole range of people – particularly artisans involved in making idols, pandals, shola ornaments for the gods and goddesses to electricians and decorators. Today these artisans are getting work for 3 to 6 months. Sometimes we are in short supply so artisans from nearby towns and villages also join in because of the demand that Durga Puja is creating. One would have thought that these traditional craftsman and artisans would lose their craft and livelihood as there would be no demand for this type of work in today's world. But it is exactly the opposite in Kolkata. Our art has survived and I can confidently say it has flourished. We have become more adventurous and each year we are making more creative, bigger…fabulous pratimas (idols). These are stories of our experimentation and innovation. Take Kumortoli, the potter's district where all the idols are made, it has become synonymous with Durga Puja in Kolkata. Today Kumortoli itself has become a place of tourist attraction … throughout the year many Indian and foreign tourists visit Kumortoli to catch a glimpse of where all the Maa Durga *pratima* are made.

Here, the narratives emphasise not only the contribution of Durga Puja to the economy of West Bengal but the kind of innovative interventions that are made despite a slowdown in the economy. Thus, the Durga Puja as a festival is reflective of the artistic experiments and innovations that are made in the face of adverse economies on the one hand. On the other, it underlines the spirit of the people and the city and its ability to bounce back in the face of all adversities, the joie de vivre that Kolkata is known for, and justifying why it has been called the "City of Joy".

Durga Puja, as Sala (2016) argues, becomes a festival as it evolves into a melting pot that stands at the intersection of culture, experience, and economy. In other words, Durga Puja undergoes a process of festivalisation, similar to how Negrier (2014) describes the processes festivalisation wherein a cultural activity that was earlier held regularly at particular times or seasons is reconfigured as a new event. Thus, it is a one-off event that is organised annually to celebrate some aspect of local culture and becomes an unusual point of convergence of people (Gibson and Stewart, 2009) Cudny (2014, p. 43) points out that festivals create in the urban landscape space for temporary cultural consumption and transforms an ordinary everyday space into a cultural space occupied for the festival. And as it is argued in this chapter, the Durga Puja emerges as a festival that becomes central to Kolkata's urban identity transforming its urbanscape during the timeframe of the Durga Puja and the branding of the city of Kolkata as a site of modern creative cultures.

Part of the city branding process entails creating the idea of discovering something new and unique that differentiates one place from another (Ashworth, 2009). In a way, city branding involves establishing a unique identity of the city. Ashworth (2009) points out three major instruments that facilitate the process of branding namely personality association, signature building, and design and event hallmarking. Interestingly, Durga Puja stands at the intersection of all the above three instruments of city branding; one, the secularisation of a religious event that creates the unique identity and persona of the city of Kolkata. Two, the organisation of the event of Durga Puja in itself at such grand and spectacular scales, and three, the pandals that form the signature buildings and design of the urban landscape of Kolkata, albeit temporarily. It brands the city as a site of not only a vibrant and culturally creative city but a city with the unique ability to give a modern twist and secular interpretation to an otherwise traditional religious practice.

Social media and the imagineering of Kolkata

The literature on city branding primarily focuses on the strategies that are adopted to promote a city and give it a competitive edge to attract resources, businesses, investors, and tourists (Cudny, 2020; Kavaratzis and Ashworth, 2005). The strategy in the branding process usually involves the "construction, communication, and management of the city's image" (Kavaratzis and Ashworth 2005, p. 507). It is this positive brand image that informs both residents and prospective visitors, tourists, investors, businesses, and resources about the city's attractiveness on the one hand (Cudny, 2019). On the other hand, they encounter the city through these brand images and make a sense of the city. In other words, the brand images influence how one evaluates and assesses the opportunities that the city has to offer (Kavaratzis and Ashworth, 2005). Thus, the process of branding is essentially a conscious selection of attributes that would produce, in most cases, a positive perception of the city making it a "conscious and planned practice of signification and representation" (Firat and Venkatesh, 1993, p. 246). The process of branding therefore articulates unique attributes of the city that acquire a meaning. Hankinson and Cowking (1993) argue that the conspicuousness of a brand is positioning and personality in comparison to other competitive brands and is primarily a combination of functionality and symbolic values.

An integral part of city branding is the articulation and circulation of the city's image and how it enables one to imagine one's possible encounter with the city. At the same time, it evokes and awakens a desire to encounter the city. The point that should be stressed is that the articulation and circulation of any city's brand have witnessed a transformative moment due to the digital revolution and the intrusion of social media in our lives.

Arko, a young Ph.D. scholar told during an interview that,

> You exist only if I google and can find you on google. Almost anything you want to know or make sense of exists on the internet. Today you are completely informed by it. So often a city appeals to me or anyone for that matter because there may be something that has gone 'viral' whether it is positive or negative. You automatically are flooded with this overload of information of different kinds…from pictures, photographs to blogs, vlogs, videos, memes you name it … it is there in social media and only a click away".

Arko's narrative is particularly important in the way he connects how one's access to social media, where information is "only a click away" has affected the circulation of brand images. Social media not only provides easy access and circulation of brand images but also makes way for it to take different forms and mediums through which brand images can be communicated in terms of "pictures, photographs to blogs, vlogs, videos, memes".

Kavaratzis (2004, p. 67) states that there are three distinct types of communication through which the city image is communicated – primary, secondary, and tertiary. Primary communication is related to the communicative effects of the city's actions where communication is not the main objective while secondary communication is the formal, intentional communication that is commonly used in marketing like advertisements, logos, graphic design, and public relations. Tertiary communication on the other hand is the uncontrolled communication that primarily takes place through word of mouth and is reinforced by media.

Social media mainly falls under the third category of communication that cannot be controlled by marketers and is self-generating in terms of consumers of the city brand, are equally producers of it and vice versa. What social media has created is a self-generating mode of not just circulation but that of articulation and production of brand images by the consumer. Thus, a simple act of taking a selfie in front of a particular monument or place, to taking pictures of that place, narrating one's experience of it, and posting them on social media platforms like Facebook, Instagram, Snapchat, Twitter works towards the production and circulation of the brand image. The question here is who is producing this brand image? Is it only the consumer, the marketer, the advertisement firm, the copywriter, the local resident, or the social media influencer, blogger, or simply a curious bystander?

Hakala (2015, p. 3) points out the participation of a wide range of people in social media and the intricate web of communication in social media, which is multifaceted, multi-step, and horizontal, resembling a web of conversations between various stakeholders and actors. Thus, social media defies the traditional means of communication in the branding process where the image and the brand are controlled by the marketer alone and consumed only by the end-user. Social media in other words have democratised and made the process of branding participatory where now the consumer with other stakeholders and actors collaborates with the marketer in co-creating the brand (Hakala, 2015). Scholars are now increasingly recognising this participatory approach in the branding process where the branding process is seen as a dialogue between the various stakeholders (Hatch and Schulz, 2010; Kavaratzis and Hatch, 2013).

Rakesh Mehta, a tourist during the Durga Pujas acknowledges,

> My interest in Kolkata was perked when one of my Bengali colleagues at the office shared a meme made by The Bong sense on FB (*Facebook*) that said something like Durga Puja is like the 7 wonders coming to the city as pandals[18]. And then my colleague showed me pics (*photographs*) of few Durga Puja pandals on Instagram and then on I was both hooked on and fascinated by it and wanting to experience it... and what an experience it has been... we got to see Maa Durga idol made out of gold. They used 50 to 60 kgs of gold to make that idol. It's unbelievable!!!

Rakesh, here points out how a meme and photographs posted on social media platforms not only piqued his interest but aroused a desire in him to seek and experience the Durga Puja in Kolkata. On close examination of the social media, we see people putting up images of the ten-day Durga Puja that chronicles every aspect of it. In fact, the social media posts and imageries begin before the festival with photographs of the idols being made, finishing touches being put to the Pratima, to the pandals being constructed that carry the tag lines of the "Maa Ashchen - The journey begins",[19] "The countdown has begun[20]" to "Work in progress... Few days left for the extravaganza, the hard work dedication of these men is irreplaceable. Their contribution towards the art and creativity of pandals makes every Durga Puja special and worth remembering among Calcuttans[21]". Series of posts[22] can be seen that Joy, a regular Instagrammer, says creates that, "Puja puja feeling". In other words, fervour and excitement that surround the Puja are produced by depicting how the city readies itself for the Pujas from the rituals of Tarpan[23] and Mahalaya[24] to the making of the Puja pandals and shopping with friends and family. And during the pujas every day, ordinary photographs in the streets of Kolkata teeming with people and enjoying and experiencing the Pujas from dhunuchi dance,[25] sindhur khela,[26] the dhakis to the unique and creative Durga Puja pandals that people explored while puja parikrama or pandal hopping during the festival[27] and transforming the city and the Durga Puja festival into an experience that is larger than life and phantasmogorical that seems to overwhelm the senses.

The city teeming with people dressed in their colourful best, the crowds on the streets, the many and spectacular pandals, the music, the lights, the food stalls together, creates an atmosphere that denotes an exuberance and joy of living that Kaushal and Newbold (2015) describe as tamasha. Joy, further points out to me that these instamoments draw out the essence of Kolkata and why it is still popularly referred to as the city of joy. He narrates,

> Take, for example, the FB post on the list of Durga Pujas to visit that went viral[28] or the photograph of the lone policeman on his bike patrolling the city

and the comments that while the rest of us are enjoying the police force has sacrificed their happiness and sleep to ensure the smooth management of the Durga Puja carnival[29], or the pics on Bijoya dashami[30], the penultimate day of the puja, the immersion of the Pratima that announces the end of the pujas and the start of the countdown for the next years' puja …you know how we keep ranting *aasche bachor abar* (once again next year) are no commercial advertisements but truly expresses the almost palpable feeling of the city, it's intense emotion and attachment to the Durga Puja Festival.

The argument here is that social media, whether it is Facebook, YouTube, or Instagram, creates an experience of real time, real place, and feel that creates a sense of authenticity. Thus, the social media imageries become difficult to separate from the real as it induces a feeling of being "there" and having a real encounter with the place in itself (Wearing et al., 2010). In many ways, the social media imagineering and representation produce an all-enveloping experience which is a highly immersive experience as well (Stevenson, 2020; Jordon 2016). Moreover, the representations on social media are shared personalised stories that are associated with their feelings making it a unique moment to be remembered and shared. It is thus able to convey the affect at the moment that can connect and involve the other viewers on social media. In other words, social media imagineering articulates not just the material spaces of the city but its symbolic spaces – of how it feels and what it means to be in that space and in that moment (Ashworth and Voogd, 1990). It is this ability of social media to produce images and imagination of a city that feels real and authentic, as in the case of Durga Puja in Kolkata, also given its self-generating and participatory mechanism makes it an effective mode of city branding.

Conclusion

"The City of Joy and Durga Puja are synonymous. This emotion and culture evoking festival have now transcended the barriers of being "just a festival". Durga Puja is the best instance of the public performance of religion and art in the city. It witnesses a celebration of craftsmanship, cross-cultural transactions, and cross-community revelry. And so, rightly and deservedly, Kolkata's Durga Puja is just a step away to get recognition in the Intangible Cultural Heritage of Humanity list of 2020 by UNESCO"[31] (Datta 2019).

The narrative published in an online webzine highlights the successful branding mechanism of Kolkata's Durga Puja that has enabled it to be recognised as intangible cultural heritage and be India's nomination to the Intangible Cultural Heritage of Humanity list for the year 2020. City branding has entirely focused on the city as an entrepreneurial one which mainly aims to improve its competitive edge over other cities in order to attract flows, both global and local, of residents, tourists, investors, businesses, capital, and other resources. The process of branding the city has not been simple but is, in fact, a complex one that has involved many stakeholders, with multifarious forms of construction, articulation, and circulation of the brand image.

Social media, as it was argued in this chapter, play a pivotal role in the manner in which the branding process takes place. In creating a distinctive and unique city brand social media creates a space for the participation of the many stakeholders, their opinions, perceptions, and experiences. But more importantly, social media has created a space through which stories that are not make-believe but real, experienced by real people in real time, and space is shared making it possible for others to contextualise and associate with the branding of the city. The real-ness of the events that gets articulated and represented in social media adds authenticity to the branding process and does not appear to be manipulated to sell the city brand. It also brings to light the intense personalisation of the branding process of cities, where the consumers co-create the brand and claim a greater stake in the branding process and at the same time infuses a sense of trust through their participation and articulation of their experiences. Much of one's experiences are mass-mediated particularly through the strong presence of social media in our everyday lives. Thus, the branding of cities inadvertently occurs through social media, where the resident or a tourist or any of the stakeholders, have an equal desire to share one's experience of the city, which is perceived as being inherently spontaneous. Thus, making the branding process authentic and trustworthy and not manipulated simply to make the city saleable.

Notes

1 https://www.youtube.com/watch?v=N8eNTFEPFsM (last accessed on 30.12.2019).
2 Pakistanis React to Ethereal Durga Pujo video https://www.youtube.com/watch?v=tjU sDGqpI14; (last accessed on 30.12.2019);
    Indonesians React to Ethereal Durga Pujo video https://www.youtube.com/watch?v =zTKoJ1FcEqg; (last accessed on 30.12.2019);
    Americans React to Ethereal Durga Pujo video https://www.youtube.com/watch?v =O0WhW9b15IY (last accessed on 30.12.2019).
3 Reaction Check on Ethereal Durga Pujo https://www.youtube.com/watch?v=N5_rfwsi 1EY (last accessed on 30.12.2019).
4 For details, see https://www.thehindu.com/news/national/other-states/burst-of-creativit y-at-kolkata-durga-pandals/article29611757.ece; (last accessed on 30.12.2019); https://www.deccanherald.com/national/kolkata-durga-puja-to-don-theme-of-fake-real-news -759530.html (last accessed on 30.12.2019); https://www.indiatoday.in/india/story/we st-bengal-durga-puja-pandal-made-of-letterboxes-invokes-nostalgia-1606324-2019-1 0-04 (last accessed on 30.12.2019).
5 Balakot airstrike refers to the airstrike carried by Indian air force targeting suspected terrorist training camps in Pakistan-occupied Kashmir. For details, see https://www.bbc .com/news/world-asia-47366718 (last accessed on 30.12.2019).
6 NRC or the National Registry of Citizenship is a registry maintained by the Government of India recording relevant information about Indian citizens. The NRC is specifically designed for the North Eastern state of Assam and was not updated till recently in 2018. The political controversy centres around the exclusion of many citizens from the NRC, and primarily those have been excluded who belong to minority religious group. For details, see https://scroll.in/article/889124/fear-mongering -by-media-politicians-over-assams-national-register-of-citizens-needs-to-stop   (last accessed on 30.12.2019).

7 See https://www.youtube.com/watch?v=1yyHNLdn2vw&t=10s (last accessed on 30.12.2019).

8 See https://www.facebook.com/photo.php?fbid=2396866897220751&set=t.18109642 50&type=3&theater (last accessed on 30.12.2019).

9 See https://www.news18.com/news/buzz/durga-puja-in-kolkata-is-not-just-about-fish-old-heritage-houses-and-dhunuchi-dance-2331557.html (last accessed on 30.12.2019).

10 For details, see https://www.theweek.in/wire-updates/international/2018/09/07/fes71 -uk-durga%20puja.html (last accessed on 30.12.2019).
   https://www.theguardian.com/travel/gallery/2018/aug/28/bengals-durga-puja-a -hindu-festival-in-full-flow-in-pictures (last accessed on 30.12.2019).

11 J.K. Rowling is the British writer who authored the Harry Potter fantasy series.

12 For details, see https://www.facebook.com/The.Educated.Moron/posts/a-beautiful -write-up-by-vir-sanghvi-on-kolkata-durga-pujawhat-pujo-means-to-a-be/149011072 4617773/; (last accessed on 30.12.2019) https://www.facebook.com/TrulyKolkata/po sts/a-man-in-delhi-once-asked-me-what-is-so-special-about-durga-puja-in-kolkata-its-/624097450976151/ last accessed on (30.12.2019).

13 For details, see http://virsanghvi.com/Article-Details.aspx?key=858 (last accessed on 30.12.2019).

14 For details, see https://www.facebook.com/subhalina.dasgupta/posts/8386210195 15455(last accessed on 30.12.2019); https://www.facebook.com/maloshree.sarkar/ posts/1518766871505673 (last accessed on 30.12.2019); https://www.facebook.com/ Simply.Shatadru/posts/10152271028174689(last accessed on 30.12.2019).

15 For details, see http://www.businessworld.in/article/The-Puja-Economy/20-09-2019-1 76487/ (last accessed on 30.12.2019); https://timesofindia.indiatimes.com/city/kolkata /not-just-fun-kolkatas-durga-puja-is-serious-business-worth-rs-15000-cr/articleshow /71549101.cms (last accessed on 30.12.2019).

16 See https://www.facebook.com/sudatta.mitra1/posts/10214667998078511 (last accessed on 30.12.2019).

17 See https://www.facebook.com/shirshendu.gayen/posts/674590672564813 (last accessed on 30.12.2019).

18 For details, see https://www.huffingtonpost.in/2016/10/10/these-durga-puja-memes-by -bong-sense-will-hit-every-bengali-righ_a_21578015/ (last assessed on 30.12.2019).

19 For details, see B3wiLH6JMaw/?igshid=14fdokec69tzn (last accessed on 30.12.2019).

20 For details, see https://www.instagram.com/p/B201pLDHV-I/?igshid=1nnwtk30gb3t2 (last accessed on 30.12.2019).

21 https://www.instagram.com/p/B1XyrmUAq_h/?igshid=2ar23vftazou (last accessed on 30.12.2019).

22 For details, see https://www.instagram.com/p/B2-r8JYAvRj/?igshid=1qei4lx88nntf; (last accessed on 30.12.2019); https://www.instagram.com/p/B3D0o7Bg0dl/?igshid =1elyyv7388rs0; (last accessed on 30.12.2019).
   https://www.instagram.com/p/B2k69Plgmeq/?igshid=16ckj22su37pr; (last accessed on 30.12.2019);
   https://www.instagram.com/p/B24WUT2nr-E/?igshid=1wgf8zli1p45; (last accessed on 30.12.2019);
   https://www.instagram.com/p/B28yYann_wf/?igshid=18oq6odmzofmq; (last accessed on 30.12.2019).

23 https://www.instagram.com/p/B28yawCHX9F/?igshid=1les7tpdu5op6 (last accessed on 30.12.2019).

24 https://www.instagram.com/p/B2-r8JYAvRj/?igshid=1k5b8f279borh (last accessed on 30.12.2019).

25 https://www.instagram.com/p/B3UmzC4HGLV/?igshid=1fobwvd7yj3ms (last accessed on 30.12.2019).

26 https://www.instagram.com/p/B3W6fR_nsGY/?igshid=16pspw8737jsp (last accessed on 30.12.2019).
27 https://www.instagram.com/p/B3TaDR_g8gA/?igshid=9eyrrpsqe53t (last accessed on 30.12.2019);
  https://www.instagram.com/p/B3QsHBxg2jO/?igshid=45ds1r1i3314 (last accessed on 30.12.2019);
  https://www.instagram.com/p/B3JgQKaH2Xf/?igshid=1h5sgjt5q44bc (last accessed on 30.12.2019);
  https://www.instagram.com/p/B3Ki4EHnWBd/?igshid=1v4wdw5d7731l2 (last accessed on 30.12.2019);
  https://www.instagram.com/p/B3KkcvOnkb8/?igshid=1aim9g0s8qrzl (last accessed on 30.12.2019);
  https://www.instagram.com/p/B4KlxI7Jvr_/?igshid=bi7pl1csx7r4 (last accessed on 30.12.2019);
  https://www.instagram.com/p/B4JfmVsJUYE/?igshid=19b3ll03curbw (last accessed on 30.12.2019).
28 https://www.facebook.com/i.am.purnendu/posts/2457891071143323 (last accessed on 30.12.2019).
29 https://www.instagram.com/p/B3lfKFagS5b/?igshid=1rfzzu9iysqc4 (last accessed on 30.12.2019).
30 https://www.instagram.com/p/B3ciLE0nujn/?igshid=j5ywm75fj0kv (last accessed on 30.12.2019).
31 https://www.whatsuplife.in/kolkata/blog/kolkata-durga-puja-unesco-world-heritage-list/ (last accessed on 30.12.2019).

Bibliography

Ashworth, G. and Voogd, H. (1990) *Selling the City: Marketing Approaches in Public Sector Urban Planning*. London: Belhaven.

Ashworth, G.J. (2009) The instruments of place branding: How is it done? *European Spatial Research and Policy*, 16(1), 9–21.

Bandhyapadhay, S. (2017) A case study of the Durga puja festival of the Bengali Hindus. *Anthropo Open Journal*, 2(1), 15–22.

Banerjee, S. (2004) *Durga Puja: Yesterday, Today and Tomorrow*. New Delhi: Rupa.

Bean, S.S. (2012) The unfired clay sculpture of Bengal in the artscape of modern South Asia. In R.C. Brown and D.S. Hutton (eds) *A Companion to Asian Art and Architecture* (pp 604–628). Malden: Blackwell Publishing Ltd.

Bennett, A. and Woodward, I. (2014) Festival spaces, identity, experience and belonging. In A. Bennett, J. Taylor, and I. Woodward (eds) *The Festivalisation of Culture* (pp 11–26). Surrey: Ashgate.

Bennett, A., Taylor, J. and Woodward, I. (eds) (2014) *The Festivalisation of Culture*. Surrey: Ashgate.

Bhattacharya, T. (2007) Tracking the goddess: Religion, community and identity in the Durga puja ceremonies of nineteenth century Calcutta. *Journal of Asian Studies*, 66 (4), 919–962.

Chakrabarti, K. (2001) *Religious Process: The Puranas and the Making of a Regional Tradition*. New Delhi: Oxford University Press.

Cudny, W. (2014) The phenomenon of festivals: Their origins, evolution and classifications. *Anthropos*, 109 (2), 640 –656.

Cudny, W. (2019) *City Branding and Promotion: The Strategic Approach*. London and New York: Routledge.

Cudny, W. ed. (2020) *Urban Events, Place Branding and Promotion*. London and New York: Routledge.

Firat, A.F. and Venkatesh, A. (1993) Postmodernity: The age of marketing. *International Journal of Research in Marketing*, 10, 227–249.

Getz, D. (1991) Assessing the economic impacts of festivals and events: Research issues. *Journal of Applied Recreation Research*, 16 (1), 61–77.

Ghosh, A. (2006) Durga puja: A consuming passion. Seminar 559 available at: https://www.india-seminar.com/2006/559/559%20anjan%20ghosh.htm (last accessed on 30.12.2019).

Gibson, C. and Stewart, A. (2009) *Reinventing Rural Places: The Extent and Impact of Festivals in Rural and Regional Australia*. Wollongong, Australia: University of Wollongong.

Hakala, U. (2015) Tracing for one voice: The 5Cs of communication in place branding. In F.M. Go, A. Lemmetyinen, and U. Hakala (eds) *Harnessing Place Branding through Cultural Entrepreneurship*. (pp. 229–242) London: Palgrave Macmillan.

Hankinson, G. and Cowking, P. (1993) *Branding in Action*. London: Mc-Graw Hill.

Hatch M.J. and Schulz, M. (2010) Towards a theory of brand co-creation with implications for brand governance. *Brand Management*, 17(8), 590–604.

Hoskins, A. (2017) *Digital Memory Studies: Media Pasts in Transition*. London and New York: Routledge.

Jordon, J. (2016) Festivalisation of cultural production: Experimentation, spectacuralisation and immersion. *ENCATC Journal of Cultural Management and Policy*, 6(1), 44–55.

Kaushal, R. and Newbold, C. (2015) Mela in the UK: A 'travelled and habituated' festival. In C. Newbold, C. Maughan, J. Jordan, and F. Bianchini (eds) *Focus on Festivals: Contemporary European Case Studies and Perspective* (pp 18–27). Oxford: Good Fellow.

Kavaratzis, M. (2004) From City marketing to city branding: Towards a theoretical framework for developing city brands. *Place Branding*, 1(1), 58–73.

Kavaratzis, M. and Ashworth, G.J. (2005) City branding: An effective assertion of identity or a transitory marketing trick? *Tijdschrift voor Econimische en Sociale Geografie*, 96(5), 506–514.

Kavaratzis, M. and Hatch, M.J. (2013) The dynamics of place brands: An identity based approach to place branding theory. *Marketing Theory*, 13, 69–86.

Mukherjee, J. (2019) Durga Puja in Kolkata is not just about fish, old heritage houses and 'Dhunuchi dance' News18 Buzz available at: https://www.news18.com/news/buzz/durga-puja-in-kolkata-is-not-just-about-fish-old-heritage-houses-and-dhunuchi-dance-2331557.html.

Negrier, E. (2014) Festivalisation: Patterns and limits. In C. Newbold, C. Maughan, J. Jordan, and F. Bianchini (eds) *Focus on Festivals: Contemporary European Case Studies and Perspective* (pp 18–27). Oxford: Good Fellow.

Ray, M. (2017) Goddess in the city: Durga pujas in contemporary Kolkata. *Modern Asian Studies*, 51(4), 1126–1164.

Sala, L. (2016) *Festivalisation: The Boom in Events*. Netherlands: Boekencoöperatie.

Sircar, J. (2011) Durga pujas as expressions of 'urban folk culture'. *Times of India*, 23 October, 2011.

Stevenson, D. (2020) Branding, promotion and the tourist city. In Z. Krajina and D. Stevenson (eds) *The Routledge Companion to Urban Media and Communication* (pp 265–273) New York and London: Routledge.

Wearing, S., Stevenson, D., and Young T. (2010) *Tourist Cultures: Identity, Place and the Traveller*. London: Sage.

https://www.youtube.com/watch?v=N8eNTFEPFsM (last accessed on 30.12.2019).

https://www.youtube.com/watch?v=tjUsDGqpI14 (last accessed on 30.12.2019).

https://www.youtube.com/watch?v=zTKoJ1FcEqg (last accessed on 30.12.2019).

https://www.youtube.com/watch?v=O0WhW9b15IY (last accessed on 30.12.2019).

https://www.youtube.com/watch?v=N5_rfwsi1EY (last accessed on 30.12.2019).

https://www.thehindu.com/news/national/other-states/burst-of-creativity-at-kolkata-durga -pandals/article29611757.ece (last accessed on 30.12.2019).

https://www.deccanherald.com/national/kolkata-durga-puja-to-don-theme-of-fake-real-ne ws-759530.html (last accessed on 30.12.2019).

https://www.indiatoday.in/india/story/west-bengal-durga-puja-pandal-made-of-letterbox es-invokes-nostalgia-1606324-2019-10-04 (last accessed on 30.12.2019).

https://www.bbc.com/news/world-asia-47366718 (last accessed on 30.12.2019).

https://scroll.in/article/889124/fear-mongering-by-media-politicians-over-assams-nati onal-register-of-citizens-needs-to-stop (last accessed on 30.12.2019).

https://www.youtube.com/watch?v=1yyHNLdn2vw&t=10s (last accessed on 30.12.2019).

https://www.facebook.com/photo.php?fbid=2396866897220751&set=t.1810964250 &type=3&theater (last accessed on 30.12.2019).

https://www.news18.com/news/buzz/durga-puja-in-kolkata-is-not-just-about-fish-old-her itage-houses-and-dhunuchi-dance-2331557.html (last accessed on 30.12.2019).

https://www.theweek.in/wire-updates/international/2018/09/07/fes71-uk-durga%20puja .html (last accessed on 30.12.2019).

https://www.theguardian.com/travel/gallery/2018/aug/28/bengals-durga-puja-a-hindu-fes tival-in-full-flow-in-pictures (last accessed on 30.12.2019).

https://www.theweek.in/wire-updates/international/2018/09/07/fes71-uk-durga%20puja .html (last accessed on 30.12.2019).

https://www.theguardian.com/travel/gallery/2018/aug/28/bengals-durga-puja-a-hindu-fes tival-in-full-flow-in-pictures (last accessed on 30.12.2019).

https://www.facebook.com/subhalina.dasgupta/posts/838621019515455 (last accessed on 30.12.2019).

https://www.facebook.com/maloshree.sarkar/posts/1518766871505673 (last accessed on 30.12.2019).

https://www.facebook.com/Simply.Shatadru/posts/10152271028174689 (last accessed on 30.12.2019).

http://www.businessworld.in/article/The-Puja-Economy/20-09-2019-176487/ (last accessed on 30.12.2019).

https://timesofindia.indiatimes.com/city/kolkata/not-just-fun-kolkatas-durga-puja-is- serious-business-worth-rs-15000-cr/articleshow/71549101.cms (last accessed on 30.12.2019).

https://www.facebook.com/sudatta.mitra1/posts/10214667998078511 (last accessed on 30.12.2019).

https://www.facebook.com/shirshendu.gayen/posts/674590672564813 (last accessed on 30.12.2019).

https://www.huffpost.com/archive/in/entry/these-durga-puja-memes-by-bong-sense-will-hit-every-bengali-righ_a_21578015 (last assessed on 30.12.2019).

B3wiLH6JMaw/?igshid=14fdokec69tzn (last accessed on 30.12.2019).

https://www.instagram.com/p/B201pLDHV-I/?igshid=1nnwtk30gb3t2 (last accessed on 30.12.2019).

https://www.instagram.com/p/B1XyrmUAq_h/?igshid=2ar23vftazou (last accessed on 30.12.2019).

https://www.instagram.com/p/B2-r8JYAvRj/?igshid=1qei4lx88nntf (last accessed on 30.12.2019).

https://www.instagram.com/p/B3D0o7Bg0dl/?igshid=1elyyv7388rs0 (last accessed on 30.12.2019).

https://www.instagram.com/p/B2k69Plgmeq/?igshid=16ckj22su37pr (last accessed on 30.12.2019).

https://www.instagram.com/p/B24WUT2nr-E/?igshid=1wgf8zli1p45 (last accessed on 30.12.2019).

https://www.instagram.com/p/B28yYann_wf/?igshid=18oq6odmzofmq (last accessed on 30.12.2019).

https://www.instagram.com/p/B28yawCHX9F/?igshid=1les7tpdu5op6 (last accessed on 30.12.2019).

https://www.instagram.com/p/B2-r8JYAvRj/?igshid=1k5b8f279borh (last accessed on 30.12.2019).

https://www.instagram.com/p/B3UmzC4HGLV/?igshid=1fobwvd7yj3ms (last accessed on 30.12.2019).

https://www.instagram.com/p/B3W6fR_nsGY/?igshid=16pspw8737jsp (last accessed on 30.12.2019).

https://www.instagram.com/p/B3TaDR_g8gA/?igshid=9eyrrpsqe53t (last accessed on 30.12.2019).

https://www.instagram.com/p/B3QsHBxg2jO/?igshid=45ds1r1i3314 (last accessed on 30.12.2019).

https://www.instagram.com/p/B3JgQKaH2Xf/?igshid=1h5sgjt5q44bc (last accessed on 30.12.2019).

https://www.instagram.com/p/B3Ki4EHnWBd/?igshid=1v4wdw5d773l2 (last accessed on 30.12.2019).

https://www.instagram.com/p/B3KkcvOnkb8/?igshid=1aim9g0s8qrzl (last accessed on 30.12.2019).

https://www.instagram.com/p/B4KlxI7Jvr_/?igshid=bi7pl1csx7r4 (last accessed on 30.12.2019).

https://www.instagram.com/p/B4JfmVsJUYE/?igshid=19b3ll03curbw (last accessed on 30.12.2019).

https://www.facebook.com/i.am.purnendu/posts/2457891071143323 (last accessed on 30.12.2019).

https://www.instagram.com/p/B3lfKFagS5b/?igshid=1rfzzu9iysqc4 (last accessed on 30.12.2019).

https://www.instagram.com/p/B3ciLE0nujn/?igshid=j5ywm75fj0kv (last accessed on 30.12.2019).

8   Film, fashion, events, and city branding of Mumbai, India

Sanjukta Sattar

Introduction

The identity of a city can be understood from various perspectives. It may be time, location, dimension, the people, and their economic, social, and cultural activities, that characterise the city. Hence, it can be said that the image of a city is multifaceted. But it is the distinctive character of the city contributing to its uniqueness, its identity, making it stand out from other urban places. The image of the city also varies with people's perception about the city which takes shape from their own experience, observation, socio-spatial interaction as well as through various activities, programs, and events which put "the city on their mental maps and enforcing the positive perception of it, whether it relates to a living, visiting or investing" (Kavaratzis and Ashworth 2005). The image is not just a physical or visual element but is also how one perceives all the components of the city reflecting how one uses and accesses the city. Hence the identity of a city may be shaped by the image, impressions iconic features, and landmark structures of the city. The city's image can be further improved, made more attractive to make it distinct from other cities and to "acquire more attention and win a competition between them" (Cudny 2019 p.25) through the process of branding. In this context, it may be said that the various events ranging from cultural and religious events, fairs, festivals to financial and business meets, conferences, and workshops, reinforce and sometimes re-position the image of the city.

Mumbai, formerly Bombay, one of the most populous and culturally diverse cities of India emerged from an industrial centre in the 20th century to a global financial hub in recent years. The city is also known as the media and entertainment capital of South Asia (Shaban 2019). It is well known for various cultural and creative pursuits among which the film and television sector has played a dominating role in creating a distinct identity of the city not only within the country but also across the globe. Mumbai houses a huge Hindi film industry, popularly known as "Bollywood", which has created a unique image of the city worldwide and is often identified as the city of "Bollywood" (Hollywood of Bombay). The city has also evolved into one of the important centres for the fashion industry which has led to the branding of Mumbai as a fashion capital too. Mumbai also hosts many other cultural festivals including music concerts, sports events, literature festivals, and multi-cultural street festivals representing the cultural vibrancy of the city. The city's image as the film and fashion and entertainment capital, cultural and creative city is also celebrated and promoted through different colourful events and activities like film festivals, award ceremonies, fashion shows, and trade shows which are organised during various times of the year. All these events have an important role to play in the place promotion and marketing endeavour to "forge distinctive images and atmospheres which act as a lure to both capital and people of the right sort (i.e. wealthy and influential)" (Harvey 1989, as quoted in Riza et al. 2012, 295) as well as provide opportunities and exposure worldwide to the budding talents.

The main objective of this chapter is to study how Mumbai is branded with the help of various events, cultural activities, and festivals that contribute to or can contribute to the city's brand and its reinforcement. The research presented here starts with studying the strengths and weakness of the place under study and continue with analysis of people's perception of the city and events under study. The chapter attempts to trace the special characteristics which are closely associated with this city and explore what has been attracting or has the potential to attract the tourists, entrepreneurs, and new residents to move to Mumbai, and how that has contributed to the creation of the city's identity and brand. The opinion and feedback obtained from the in-depth interviews conducted among organisers, participants and audience of events, tourists, various professionals (e.g. journalists, actors, models, event organisers, business executives, traders, social workers academicians), students, and resident population, has helped in understanding people's perception of the city of Mumbai, and what makes this city distinct from other urban places in the country. Taking a clue from the interviews and discussions and intensive web-search, the various events, especially those which are exclusive to Mumbai were identified. The details of these events were investigated to understand and discuss the potential and effectiveness of these events, activities, and festivals in presenting Mumbai itself as a brand globally. Also, various reports, websites, and other research publications were referred for procuring data on types of events and other related information.

The chapter is divided into five sections. After this introductory section, section two attempts to conceptualise the city branding, while section three examines Mumbai's identity through people's perception. Section four discusses Mumbai as an entertainment capital, cultural centre, and a multicultural city and events reflecting these identities; the last section is the conclusion.

Conceptualising city branding

"City branding is the process of branding city or place and is called as geobranding, place branding" (Malvika 2018). This process involves two main components which include place-making or city building, i.e., making a specific place more attractive and place or city branding, i.e., promoting a place (Anholt 2008; Avraham and Ketter, 2008, Kavaratzis, 2004, Jojic 2018). Cities are hubs of economic, technological, social, and cultural opportunities and "act as self - perpetuating engines attracting even more people" (Kourtit and Nijkamp 2015, p.6). Hence, a city brand aims at developing "a strong 'umbrella' brand that is coherent across a range of different areas of activity with different target audiences, whilst at the same time enabling sector-specific brand communications to be created" (Dinnie 2011, p.5). A city is simultaneously a place where people live and visit, and also a place of opportunities attracting investments. The target audiences range from the city's current and future residents to tourists, potential investors from outside, and other internal stakeholders like creative entrepreneurs, event organisers, and urban development departments. With such a diverse group of stakeholders, the images held about the city are equally varied. A city is generally branded by those features which are very iconic for the city and presenting a very positive image of the city across multiple audiences. This is necessary so that the "cities, therefore, seek to market, represent themselves as centres of creativity, innovation and culture in the belief that this will give them a competitive edge" (Uysal 2013, 223). This also enables cities to "attract tourists, immigrants, and prospective investors" (Cudny 2020, 1), which contribute to the health, sustainability, and overall development of the city.

City branding is, therefore, a form of place management with the objective of changing or improving the way places are perceived generally or by some particular groups of people thus creating a specific image and identity of the city. This can also happen with urban renewal which "includes the creation of a new identity with its own experiential value, which is profoundly original and impossible to copy" (Uysal 2013, 223), touching upon aspects as "structure, programming, functions, the sort of actions and activities that characterize the image of the city, events and in the last resort the chemistry of the people who operate there" (Florian 2002). In fact, "City branding has become a recognized urban policy worldwide" (Uysal 2013, 223).

To enhance the appeal of a city and to increase its competitive edge, presently many urban authorities have begun to adopt the branding process as part of city marketing and urban development (Riza 2015). While competing for investments brand reputation plays a significant role as "brand is both a lens through which information is viewed and a decision criterion" (Middleton 2011, 15). Also as the competition between the cities intensifies in terms of talent attraction, tourism promotion, attraction to entrepreneurs, investment attraction, competition for being the venue for sports and cultural events, successful and strong branding becomes vital. It can contribute to the sustainability of the city in terms of living and liveable environment by creating a strong city brand through implementing environment-friendly measures benefiting the residents and visitors to the city.

Branding of cities is done in various ways. It involves implementing coherent strategies concerning managing the resources, reputation, and image (Dinnie 2011, 3). How the target audience can be reached creatively and convincingly and how the city's image is conveyed with credibility need to be planned intelligently. The image communication is done in many ways. It may be through the events and specially organised activities and programs, also through advertising, public relations, usage of logo, slogans, graphic designs, and installations.

The image of the city is also communicated through word-of-mouth communication or textual expressions of the experiences of the visitors and residents and personal opinions of the place. Communicating of images of the city, brand building, and brand management may be partly through traditional channels but also through the plethora of digital media that are now available (Dinnie 2011, 5).

Presently, the events-centred approach to city branding is practiced by many cities as one of the ways by which brand values and positive attributes can be reinforced through the association with special events and programmes. For example, Sydney's city branding through global events like 2000 Olympics, 2003 Rugby World Cup, World Youth Day in 2008, as well as Sydney-specific annual events and festivals "'Festivals of Light, Music and Ideas' which has been custom-built to showcase Sydney's special creativity" (Parmenter 2011). Branding of Athens by hosting mega-events like 2004 Olympic games, branding of Budapest as a "Festival City" by holding fairs and seasonal festivals, to name a few (Fola 2011, Szondi 2011).

In the case of some cities "hosting of mega-events may act as the starting point of city branding" (Dinnie 2011, 96). Events planned for branding the neighbourhoods within the city or the city as a whole include a festival, sports events, art performances and bundled culture and art experiences. The events are customised according to the city's background to promote the hallmark features which brand the city. The appeal of the events may vary across the different target audiences or stakeholders. The events can be categorised by their themes each portraying and promoting the different noticeable attributes of the city. Again there are some events which "help unite a whole city" (Dinnie 2011, 96). Events are also organised to reposition the image of the city by updating the contemporary developments as well as by leveraging those features which were yet to be recognised as one that can earn credit to the city and contribute to its brandings. If the branding of the city is already strong, it can attract prominent events to the city which in turn will further enhance the image of the city.

Mumbai and its identity

Mumbai – presentation of the city

Mumbai, the capital city of the state of Maharashtra located in the south-western part of the country is a principal port on the Arabian Sea. This is India's most populous city with a population of 12.44 million (Census of India 2011) and is also the city that is highly cosmopolitan and rich in cultural diversity. The city developed from seven islands which was originally the home to Koli communities who were fishermen by livelihoods. For years, these islands were under the control of native rulers until they were taken over by the Portuguese who had to give away these islands to the East India Company as a part of the dowry of the Portuguese princess who married Charles II of England.

Over time, with the building of major roads, the introduction of railways, the establishment of cotton mills, land reclamations to create more space to accommodate the increasing population, and advanced infrastructure development, the port city grew in size and importance. It developed into an industrial and significant commercial centre of the country and a global financial hub. The city hosts India's stock market, leading business houses, and regional headquarters of many multinational corporations.

Mumbai is also the centre of the entertainment industry. This adds another dimension to the city's identity – Mumbai as the media and entertainment capital of the country. The city is well known for various cultural and creative pursuits among which the film and television sector have played dominating roles in creating a distinct identity of Mumbai not only within the country but also across the globe. Mumbai houses a huge Hindi film industry, popularly known as "Bollywood", which has created a unique image of the city worldwide and is often identified as the city of "Bollywood" (Hollywood of Bombay). Moreover, as the world of cinema is intertwined with the world of fashion and Bollywood as an early trendsetter in fashion Mumbai emerged as a creative hub for a number of fashion designers who started experimenting with film fashion which in the later stage was led to be developed into the mass market. Soon Mumbai became an important centre of fashion designing and manufacturing in India. This city is today also a popular tourist destination. The growing number of tourists stopping over in Mumbai has made it the only Indian city to enter the Asia Pacific list of top 10 destination cities (Narayan 2016).

People's perception and Mumbai's identity

"Urban identity is created through a complex of interactions changing through space and time among the people who inhabit the geopolitical space a city occupies, the agencies which operate and the political leadership which evolves in and around it" (Prakash 1993, 2119). The idea of what identifies a city can vary from person to person. How are cities perceived includes place attachment and place meaning (Kudryavtsev et al. 2012). People's attachment and the perception of the city evolve through their experience in the city and what symbolic meaning they make out from it. The city's image serves as a means for city residents to identify with their city.

To the tourists, Mumbai is the city of Bollywood, a city that never sleeps. According to David, a British tourist, who is amazed by the rhythm and spirit of the city, "Mumbai is a 24/ 7 city". Mumbai's identity as a commercial and financial hub, entertainment capital, fashion hub, cultural capital, and a tourist destination gets validated by the perception of the "Mumbaikars" or the residents of Mumbai as well as by the tourists visiting the city (Figure 8.1). The city is also known for being a popular shopping destination, an additional attraction to the tourists. The availability of a variety of cuisines it offers ranging from continental dishes to the local Maharashtrian fares and dishes from other parts of India, adds to the multicultural character of the city.

Another unique feature of the city is that here both occidental and oriental cultural traits characterise Mumbai's cultural space. "It is the most western of all India cities, and the most Indian of all western cities" (Harris as in Prakash

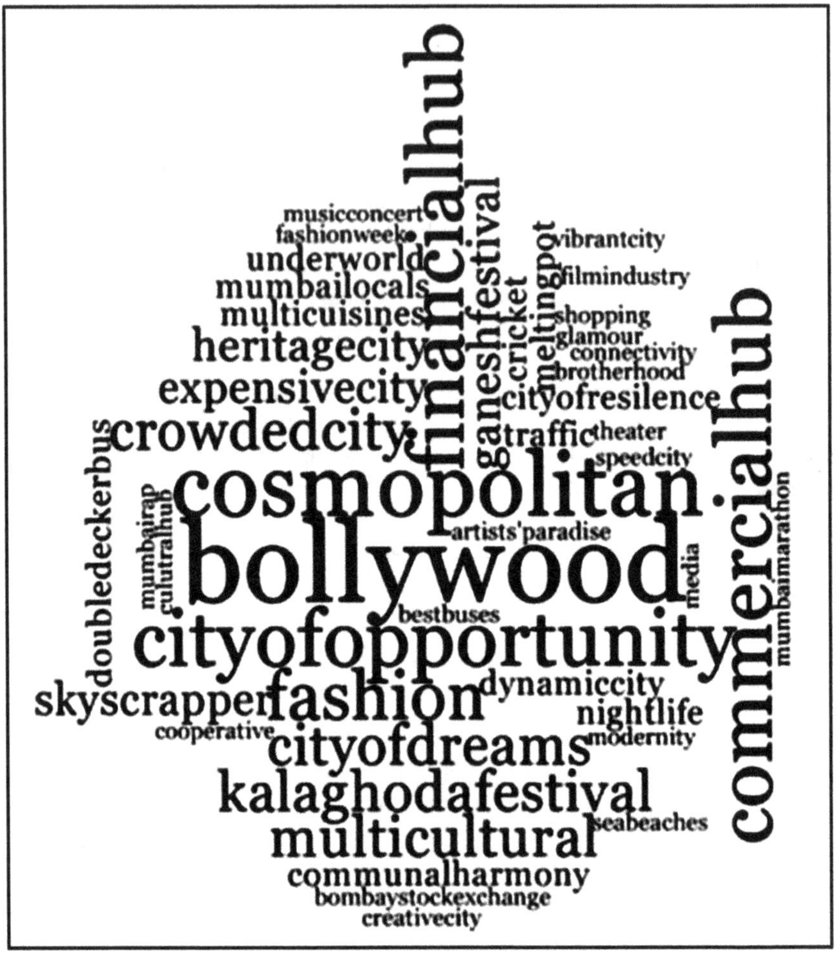

Figure 8.1 People's perception of Mumbai. Source: Based on data from the primary survey (30 respondents).

1993, 2120). Mumbai is a city that is always bustling with activities, a city where there is no dearth of work for those who do not want to remain idle but toil hard to fulfill their dreams. As mentioned by many, Mumbai is the city of dreams or *mayanagari*. "Mumbai always has been the abode not just for the people in search of livelihood but also those with a fire in their belly and dreams in their eyes" mentions a professor of media studies, who was born and brought up in Mumbai, and has observed since her childhood how the city attracts people searching for a better life. People move into this city with the hope of a successful career in the film and television industry. The flourishing arts scene and Bollywood, the Hindi Film industry, are among the major sources of attraction. Some of the newcomers may have come to the city to start their business ventures, with the aspiration of reaching a commendable position in the business world. "Bombay is a bird of gold" (Mehta 2006, 18), it is a city full of opportunities. It is believed that "nobody starves to death in Mumbai" (Mehta 2006, 18).

The city is also known for its disadvantages and darker sides like the underworld, blasts, overcrowding, high living costs, and slums. But the resilient spirit which characterises the city mobilises the resident population to overcome the difficulties, survive, and move on. This makes Mumbai a "dynamic city", a "resilient city". According to an artist based in this city, Mumbai means crowd, traffic, noise, art, cinema, and lots of business which makes the city quite similar to NYC. The city is home to diverse ethnic groups who live side by side, supporting and sharing each other's woes and worries as well as happy moments. The city is identified as a cosmopolitan centre where the residents participate and celebrate each other's festivals. This tells us why Mumbai is quoted as an "inclusive city", a "multicultural city" and a cultural melting pot. Probably this is why, Ajay who has been living in Mumbai for the last 10 years, considers Mumbai as the city where people from all parts of India aspire to come, work, stay, and live. The city, which is famous for its liberal social environment and industrial capitalism is also extremely accepting of all kinds of business and careers and is having a conducive environment to start something new and being innovative.

Mumbai as an entertainment capital, cultural center, and a multicultural city

Mumbai is well known as the entertainment capital of India. This city since long has been one of the prime centres of art and culture in the country. Mumbai is also known for its theatre culture since the 19th and 20th centuries though it did not travel beyond the national borders due to the limited mobility of the performers. Even within the country, it could appeal to only a particular section of the population due to language barriers. Literature too could penetrate a selected section of the population due to low literacy rates and limited educational attainment. But it was a cinema which overcame the limitations like restricted mobility and low literacy rate and instead reached and satisfied a much larger section of the population as a major source of entertainment. It became the most popular form of mass media. With the introduction of the television industry cinemas reached still a larger number of households. Bombay, currently Mumbai, became the growth centre of the cinema industry mainly because of the influence of the Parsi theatre artists who were based in this city The earliest Indian films in Hindi were inspired by the Parsi plays which were characterised by the blend of realism with fantasy, music accompanied by dance, narrative with spectacle and dialogue with stage presentation (Gokulsing and Dissnayake 2004, 98). Soon Bombay became the centre of growth of several filmmakers, artists, musicians, and the technical staff. The city evolved from an incubation hub into a full-fledged film and television industry. People from all corners of the country and various cultural backgrounds were significantly attracted to the city by the flourishing film and television industry in addition to the other economic opportunities found in the city. This contributed considerably to the cultural diversity of contemporary Mumbai.

Events promoting film and fashion

The Bombay film industry evolved into a significant global media industry "Bollywood", which is a multi-million dollar Indian film industry known world-wide. The term "Bollywood" was apparently coined by Amit Khanna, a former film lyricist who turned the head of a film production company as the Hindi film industry was based out of erstwhile Bombay (Free Press Journal 2018). The Bollywood movies have created an identity for themselves and are well-acclaimed even globally. Popular Hollywood actor Brand Pitt once admitted in an interview with the Indo-Asian News Service that he would enjoy working in a Bollywood movie as "there is so much drama and color in the films there" (Statista Research Department 2018).

Mumbai being one of the richest cities in the country and having better infra-structural facilities can attract resources from other parts of the country even from abroad to be invested in the making of films, which led to the development of the film industry in Mumbai. This multi-million dollar Indian film industry was formally granted industry status in 1998 and is "other than Hollywood only one media industry that has gone global" (Bravo 2014). The three major aspects emphasised by Punathambekar (2013) that encouraged Bollywood's development include:

> i) India's desire to reconfigure its national space and rebuild its relationship with Indian diaspora, ii) its dream of refashioning Mumbai as a global city in an era of economic and cultural globalization, iii) the expansion and con-vergence of Indian media, including television, radio, film, Internet, mobile phones, and advertising. The blend of these elements allowed Bollywood's rise.
>
> (Bravo 2014)

Also, new marketing strategies and promotions of Bollywood through events and celebrations were aimed to expand its national audience and increase popularity and demand among the Indian diaspora and in the global market. Events com-memorating the contribution of celebrities from Bollywood, award ceremonies, promotions and launching ceremonies of films, Bollywood Musical events, event tracing the evolution of Bollywood, are some of the various types of events related to Mumbai's Hindi film industry, which are held all through the year, further reinforcing Mumbai's identity as film capital. Mumbai had been the destination of the 20th International Indian Film Academy Awards (IIFA), 2019 ceremony, a significant addition to list of glorious events held in Mumbai reinforcing the city's identity as a hub of the film industry. Mumbai is the only city in India where KASHISH Mumbai International Queer Film Festival is held annually since 2010. This festival attempts "to encourage greater visibility of Indian and international queer cinema among both queer and mainstream audiences" (KASHISH 2021). It has also attracted many Bollywood celebrities, filmmakers, critics, and academicians from India and abroad. This event which is covered extensively by both national and international press represents the image of Mumbai as a liberal city.

Another film event that adds to Mumbai's distinction as a film and entertainment capital is Jio MAMI Mumbai Film Festival organised annually by the Mumbai Academy of Moving Image (MAMI) to provide an international platform for showcasing films from different parts of the country. Mumbai also hosts events of different nature like screening of certain theme-based films followed by panel discussions and critical appreciation of films are organised providing food for thought to the intellectual crowd from and outside Mumbai. Events are also organised for the promotion and marketing of films for stretching the market reach and increasing the revenue earned.

Bollywood special tours like Mumbai Film City Tours, Bollywood Studio Tour, Experience Live Show Tour, are organised to provide an experience of the art of Bollywood studio film shooting to the tourists as well as the resident population of the city. As a part of the film city tours, the participants are taken around a Bollywood studio, visiting the various sets used in movies and also serials, if lucky the visitors may see live shooting from a distance, witnessing how work goes on behind the scene and various other activities which are part of the moviemaking process. The tour may also include driving past the residences of popular Bollywood artists and visit the Bollywood museum where one can try wearing some famous Bollywood outfits, watching and participating in Live Bollywood Dance and Musical Shows. These provide a perfect introduction of Mumbai as the "film capital", Bollywood history, and a total experience of the unique culture of Bollywood which makes one understand how Mumbai and Bollywood go hand-in-hand.

The city has also evolved into one of the important centres for the fashion industry which has to lend the identity of fashion capital to Mumbai. It is to be noted that in the history of Indian fashion an early trendsetter was Bollywood. The fashion trend set by Mumbai's Hindi film industry is a vibrant example of the integration of film and fashion which has added a competitive edge to the city especially in attracting entrepreneurs, designers, artists, models to the city of opportunities. Komal Nahta, who is a film trade analyst, magazine publisher, and TV show host believes that "Bollywood costumes are great blueprints for real-life occasion clothing" (Kapoor 2017). There is a noticeable influence of the Bollywood fashions on the regular wear of the people as well as the wedding apparels and some of the styles from particular films "go on to become iconic trendsetters for an entire generation" (Kapoor 2017).

The outcome of the integration of Bollywood and fashion is #BollywoodFashion or #BollywoodStyle which contributes to the branding of Mumbai as a place of origin of a fashion style *Bollywood fashion*. There are many reputed fashion designers based in the city who has been designing for Indian film celebrities as well as for important personalities across the globe thus not limiting their sphere of work within Bollywood only. To give international exposure, especially to the upcoming fashion designers a number of events are organised in Mumbai. The Lakme Fashion Week (LFW) is one such fashion event that has managed to integrate Mumbai and India into the global fashion network. This event is held to give impetus to the business of fashion in India by catering to the requirements and demand of the fraternity as well as keeping it at par with international standards. It provides the platform to showcase the best Indian and International fashion talent to an audience comprising of buyers, media, and influential personalities from the global fashion industry (Chatterjee 2011).

Bombay Times Fashion Week is an event that provides opportunities to the debutant designers, the students of fashion designing to showcase their exclusive designs along with the veteran designers who are celebrities in the world of fashion India. As mentioned by a model and actor from Mumbai, the events drew the attention of many foreign directors and designers who are investing in the Indian film and fashion industry and have also attracted international as well as domestic tourists to the city. Mumbai being a significant commercial hub has set the ground suitable for holding several trade shows to feature across the world domestically manufactured garments and those designed by upcoming designers from all over the country even for special occasions like weddings. Thus Mumbai is the favoured destination for holding reputed fashion shows as well as trade shows of apparel and garments as well as beauty and fashion products (Table 8.1.).

Music and musical events

Mumbai is also a hub for music lovers and one of the important centres of the music industry. The city is also known to hold many music festivals which attracts music enthusiasts even from outside the state. The music festivals held in Mumbai cater to demand from all cross sections of the population belonging to different age groups and followers of different types and different schools of music. For those who are passionate about Western classical *Arties's Festival India* is a big attraction. This festival, in collaboration with National Centre for Performing Art, features Artie's Chamber Orchestra, performing symphony music specially arranged for chamber orchestra. Asia's biggest Blues music festival *Mahindra Blues* which takes place in Mumbai every February has been promoting the local blues community. Budding blues artists from India get an opportunity to perform on the same platform with the established blues artists from across the world. Tourists attending this festival also get a chance to taste local cuisine and experience the bustling nightlife of Mumbai. *Mood Indigo* is an annual music festival organised by the Indian Institute of Technology, Bombay, and is one of the oldest music festivals in the country. This festival which attracts students from all over India is known for inviting renowned international and national bands as well as for creating awareness about various social issues and facilitating participation in humanitarian acts like blood donation camps organised simultaneously.

Every year a number of musical events featuring performances by international celebrities like Justin Bieber, Khalid, and musical bands like Irish Rock band U2 are organised thus adding Mumbai to the list of venues of the musical

Table 8.1 Examples of significant fashion events and apparel and clothing trade shows in Mumbai

Event	Type	Frequency	Types of visitors	Aim/purpose
Lakme Fashion Week	Fashion event	Bi-annual	Designers, models, make -up artists, film celebrities	Launching the careers of designers and integrating India in the global fashion map
Bombay Times Fashion Week	Fashion event	Annual	Designers, Film celebrities, models, fashion stylists, make-up artists, students	Providing platforms to showcase designs by upcoming designers and students as well as veteran designers
India Fashion Forum	Trade show–Apparel & clothing, fashion & beauty	Annual	Owner of fashion boutiques, stylists, proprietor, director, managing director, manager of designing houses, event executives and CEO of event management houses	Exhibition and conference for fashioning Indian retail
Design Inc	Trade Show–apparel clothing, fashion & beauty	Bi-annual	Fashion designer, company owner, partner, directors, proprietors	Exhibiting work of selected designers from across India displaying fresh collections and latest trends in wedding trousseau wear, holiday and western wear, jewellery, and accessories.
Fashion Tribe Show	Trade show	Annual	Owners, proprietors, fashion designer, CEO, managing directors	Is a platform to see the latest designs

(*Continued*)

Table 8.1 (Continued)

Event	Type	Frequency	Types of visitors	Aim/purpose
National Garment Fair	Apparel Trade show	Bi-annual	Owners, proprietors, directors	Aims to feature the domestically manufactured garment industry across the world.
Parineeti – The royal Edit	Trade show – apparel, beauty and fashion	Quarterly	Owners, proprietors, directors, business partners, fashion designer	Meant for wedding shopping or lifestyle shopping, displaying Indian wear, Indo-western wear, trousseau, jewellery, footwear, etc.
70th Garment Show of India	Trade show of apparel and clothing	Bi-annual	Manufacturers, retailers, wholesalers, distributors, traders, retail chains	The mega platform for the domestic garment industry for networking, and to meet potential buyers and suppliers.
Wedding Asia	Trade show for the wedding planners	Annual	Wedding professionals and planners, clients	Provides the attendees the opportunity to meet with wedding professionals and planners for planning a perfect wedding including specially designed dresses
Bridal Asia, Mumbai	Trade Show for apparel, clothing, beauty and fashion	Annual	Fashion designers, proprietors, clients	The event gives a platform to the newcomers as well as the established professionals to help build their contacts and make their design reachable to a wider audience

(*Continued*)

Table 8.1 (Continued)

Event	Type	Frequency	Types of visitors	Aim/purpose
India International Beauty Fair	Trade show, fashion show, conferences, product launch, award shows	Quarterly	Owners, directors, proprietors, makeup artists, managers	Provides an international platform to showcase the beauty and wellness trends, brands, beauty services, and solutions.

Source: Author's compilation based on data obtained from various web sources: https://10times.com/
mumbai-in/fashion-accessories
https://timesofindia.indiatimes.com/life-style/fashion/shows/highlights-of-bombay-times-fashion
-week-2019/articleshow/68547450.cms
https://www.thehindubusinessline.com/news/variety/Lakmé-Fashion-Week-puts-the-spotlight-on
-Indian-designers/article20112381.ece.

event at the international level. The city also hosts "Hindustani" (Indian) classical music festivals like *Gunidas Sangeet Sammelan* and *Banganga* which is a two-day Indian classical music festival. *Bollywood* music or the Hindi film music has much popularity even across the national border and events like *Bollywood Music Project* showcasing experimental Bollywood music along with thematic decor and installations and multi-genre Bollywood music programming is organised in the city. It is claimed to be Asia's largest Bollywood music festival featuring the best music from the Hindi films presented by some of the popular and established as well as upcoming playback artists. In the words of a renowned Bollywood music composer

> Bollywood music has always been globally recognized and acknowledged on a large scale. In India, music festivals are becoming an important part of the music culture and it's great to see how Bollywood Music Project has created a community of music lovers that lives and breathes Bollywood.
> (http://everythingexperiential.businessworld.in/art
> icle/Amit-Trivedi-and-Vishal-Bhardwaj-to-enth
> rall-the-audience-at-Bollywood-Music-Project-5-0
> /18-11-2019-179141/)

Throughout the year various other Bollywood music events and festivals like *Golden Memories of Rafi Saheb, Timeless Classic – Geeta Dutt, Asha Bhosale Live, Kishor Kumar and Rajesh Khanna, Arijit Singh Live*, etc., remembering the legendary playback singers from Hindi film industry *Bollywood* and celebrating the music from the past as well as present are among the list of much-awaited events that adds to Mumbai's identity as the entertainment capital.

Festivals and events presenting Mumbai as a city of multiculturalism and multiple identities

Mumbai is known for its cosmopolitan culture and this has been expressed through the various events and festivals held regularly. The city is known for its grand celebration of religious festivals as well as for its very special street festival of art and cultural heritage and for the exclusive sports events which draw sports enthusiasts from far and near.

Among the various religious festivals, Mumbai is known for the celebration of the Ganapati festival. It is the festival of Ganesha, one of the popular deities in India with an elephant head, and is a symbol of wisdom and good luck. Though this festival is celebrated all over Maharashtra and also in some other parts of the country, the grandeur, pomp, and enthusiasm that is seen in the celebration in Mumbai make it very iconic to this city. In 1893, Indian freedom fighter and social reformist, Lokmanya Bal Gangadhar Tilak, transformed this annual domestic festival in Maharashtra into a large public event (Maharashtra Tourism Development Corporation 2015) to bring together people of all faith and unite them to fight against the British colonial rule. Henceforth it became one of Mumbai's grandest festivals celebrated by everyone regardless of their faith. Till today, as encouraged by Lokmanya Tilak, gigantic images of Lord Ganesha are installed in the public pavilions which are gorgeously decorated. After 10 days, idols are taken out in elaborate procession for immersion in the sea or in water tanks specially designated for that. The locals, as well as tourists from all over the country and abroad, visit Mumbai during this festival to experience the grand celebration.

Mount Mary Fair and Fair at Mahim Dargah (shrine) are also attended by a large number of people from across the society. Mount Mary Fair is an 8-day event to celebrate the birth of the Virgin Mary. Mahim Mela (Fair) is held annually in honour of the Sufi saint Makhdoom Fakih Ali Mahimi and continues for 10 days during December. It is noticeable that these fairs are attended by people of all faith with equal enthusiasm thus reinforcing the claim of Mumbai as a multicultural city.

Kala Ghoda Art festival, one of the most prominent cultural events in Mumbai which is held every year at the beginning of February attracts art enthusiasts from the city as well as outside to South Mumbai. This festival is organised by the Kala Ghoda Association which was formed on 30 October 1998 with the aim of maintenance and preservation of the iconic structures and to refurbish the art district of South Mumbai. As part of this endeavour, since 1999, an art festival is being annually organised mainly to draw attention to this area's art and architectural heritage, to "create and spread multi-cultural awareness through platforms like festivals and events especially amongst those who have little opportunity to access or be exposed to culture" (Kala Ghoda Association 2016) and also to generate funds for sponsoring projects for the conservation of the area. The area called Kala Ghoda (Black Horse) in South Mumbai where this festival is held got its name from the old equestrian statue of King Edward VII which was placed at the nodal junction and that was how the place was colloquially referred to. Though presently in place of the old equestrian statue a new installation of a black stone horse stands representing the name and recalling the history of the precinct. The "Kala Ghoda precinct has an existing critical mass of art galleries, museums, and cultural spaces unrivalled in all of India, and perhaps comparable to art districts present in other parts of the world" (Kala Ghoda Association 2016).

During this art and culture festival, which spans over nine days, varieties of engaging and interesting programmes are organised which includes theatre, film screening, dance, music, literature, traditional and contemporary display of art and art workshops, heritage walk tours to provide participants an insight into the city's heritage, display and sale of handloom and handicraft products by craftsmen coming from all corners of the country, and also special culinary sections holding live food demonstrations on varieties of cuisines by well-known chefs. This street festival where many artists, performers, and craftsperson come together attracts huge crowds from near and far. Many tourists not just from the city but from all over the country, as well from across the world, plan their visit during this time to get an experience of *Kala Ghoda festival*. Moreover, in addition to the overwhelming impact of the various activities during this festival, this particular area having the concentrated collection of heritage structures and historical buildings, which are the iconic city landmarks, leaves behind a deep impression on the tourists and the visitors which they take back as the memory of this city. The uniqueness of this art and culture event lies in the fact that it integrates and amalgamates modernity with tradition, contemporary with heritage, which very much characterises Mumbai culture.

Mumbai Litfest also known as *Tata Literature Live!* is one of the largest international literary festivals featuring authors, poets, thought leaders. The event is thronged by book lovers where they can meet various prominent authors and get their books autographed. This festival which has been praised by reputed American writer Meg Rosoff as "the loveliest literature festival in the world" (TATA Literature Live 2019), includes bold discussions, conversations, educative workshops, exciting performances, talks, debates related to literature, as well as other contemporary topics. They encompass agrarian distress, climate change, water crisis, mental health, economic scenario, arts and culture, and even cricket. This is an intellectually stimulating event for people of all ages and praised as has put Mumbai's name on the world map of literary festivals.

Tata Mumbai Marathon is another event that glorifies Mumbai's name in the world map of sports events, especially distance running. This is amongst the top 10 marathons in the world and is the largest mass sporting event in Asia which attracts participants from across the world of varying abilities, caste, creed, and social strata. Since 2004 this event is being held in Mumbai on every third Sunday of January every year. This event represents not only the biggest participative sport in the country but also the single largest fund-raising platform for Civil Society Organisations, which has raised approximately INR 266 crores (about US $38 million at current prices) benefitting over 700 NGOs (Tata Mumbai Marathon (TMM) n.d.) working on various social issues for ushering in a better tomorrow.

Hence this event definitely adds to Mumbai's identity as a venue for the international sports event.

Conclusions

Mumbai among others has been known as an entertainment capital and it is solely attributed to the existence of Bollywood, the Hindi film industry. After Hollywood, this is the only film industry in the world that has global outreach creating a brand globally. Bollywood has created a distinctive image of Mumbai and contributed to refashioning Mumbai as a city with global recognition. So it may be said that Mumbai is branded as the city of Bollywood. Various events related to the Hindi film industry that are organised are mainly for the promotion and presentation of the latest achievements of the industry. These events also support the development of those sectors like fashion and music which are intertwined with Bollywood. Thus these events indirectly contribute to the branding of Mumbai as an entertainment capital that revolves around the existence of Bollywood.

The city is also known for various cultural, sports, and creative pursuits represented through various events like the International Literature Festival, Mumbai Marathon, music concerts, exhibitions which have added to Mumbai's identity as a centre of sports and cultural activities. But Kala Ghoda Art and Culture Festival is one that may be considered to be a landmark festival of Mumbai, as it portrays the cultural heritage and the multicultural ethos of this city which makes Mumbai distinct from others. As seen in most situations, the most desirable branding instruments are hallmark events linked to signature districts (Kołtun 2020). Kala Ghoda Art Festival is also an example of a hallmark event coupled with the signature art district of South Mumbai. These events definitely have an important role to play in the place promotion and marketing endeavour by adding a competitive edge to the city in terms of attracting investments, sponsors, tourists, new talents, and entrepreneurs.

References

Anholt, S. (2008) Place branding: Is it marketing, or isn't it ?". *Place Branding and Public Diplomacy*, 4 (1), 1–6.
Avraham, E. & Ketter, E. (2008) *Media Strategies for Marketing Places in Crisis: Improving the Image of Cities, Countries and Tourist Destinations* (vol 1, pp. 222–243.). Oxford: Butterworth Heinemann.
Bravo, M.M.L. (2014) Book review (From Bombay to Bollywood: The Making of a Global Media Industry by Aswin Punathambekar, New York University Press, 2013). *International Journal of Communication*, 8, 161–164.
Census of India (2011) Mumbai (greater Mumbai) city census 2011 data, available at: https://www.census2011.co.in/census/city/365-mumbai.html (accessed on 3rd December 2019).
Chatterjee, P. (2011) Lakme fashion week puts the spotlight on Indian designers. *The Hindu, Business Line*, March 24, 2011, available at: https://www.thehindubusinessli

ne.com/news/variety/Lakmé-Fashion-Week-puts-the-spotlight-on-Indian-designers/ar ticle20112381.ece (accessed on 4th December 2019).

Cudny, W. (2019) *City Branding and Promotion: The Strategic Approach*. New York: Routledge.

Cudny, W. ed. (2020) *Urban Events, Place Branding and Promotion Place Event Marketing*. London: Routledge.

Dinnie, K. (2011) Introduction to city branding. In: Keith Dinnie (ed) *City Branding: Thoery and Cases* (pp.3–7). Basingstoke, Hampshire, UK: Palgrave Macmillan.

Everything Experiential. (2019) Amit Trivedi and Vishal Bhardwaj to enthral the audience at Bollywood Music Project 5.0. *EE News Desk*, November 2019, available at: http: //everythingexperiential.businessworld.in/article/Amit-Trivedi-and-Vishal-Bhardw aj-to-enthrall-the-audience-at-Bollywood-Music-Project-5-0/18-11-2019-179141/ (accessed on 3rd December 2019).

Florian, B. (2002) The city as a brand: Orchestrating a unique experience. In: Hauben T, Vermeulen M and Patteeuw V (eds) *City branding: Image Building and Building Images*. Rotterdam: NAI Uitgevers.

Fola, M. (2011) Athens city branding and the 2004 Olympic games. In: Keith Dinnie (ed) *City Branding: Thoery and Cases* (pp. 3–7). Basingstoke, Hampshire, UK: Palgrave Macmillan.

Free Press Journal (2018) Mumbai and Bollywood! Inseparable bond, with entertainment guaranteed. *Entertainment*, July 27, 2018, available at: https://www.freepressjournal.in /entertainment/mumbai-and-bollywood-inseparable-bond-with-entertainment-guarant eed (accessed on 3rd December 2019).

Gokulsing, K.M. & Dissnayake, W. (2004) *Indian Popular Cinema: A Narrative of Cultural Change*. Staffordshire: Trentham Books.

Harvey, D. (1989) *The Condition of Postmodernity: An Enquiry into the Origins of Culture*. Oxford: Blackwell.

Jojic, S. (2018) City branding and the tourist gaze: City branding for tourism development. *European Journal of Social Science Education and Research*, 5(3), 150–160.

Kala Ghoda Association. (2016) Kala Ghoda Arts Festival, available at: http://www.kala ghodaassociation.com/about-kga.html# (accessed on 4th December 2019).

Kapoor, A. (2017) Indian fashion's obsession with bollywood. *Open: Fashion*, March 2017, available at: https://openthemagazine.com/features/fashion/indian-fashions-obse ssion-with-bollywood/ (accessed on 3rd December 2019).

KASHISH (2021) KASHISH Mumbai international queer film festival, available at: https://mumbaiqueerfest.com/vision/ (accessed on 20th April 2021).

Kavaratzis, M. (2004) From city marketing to city branding: Towards a theoretical framework for developing city brands. *Place Branding*, 1(1), 58–73.

Kavaratzis, M.,& Ashworth, G. (2005) City Branding : An effective assertion of identity or a transitory marketing trick?, *Tijdschrift voor Economiche en Sociale Geografie*, 1(1), 506–514.

Kołtun, A. (2020) In Cudny Waldemar (ed) *Urban Events, Place Branding and Promotion Place Event Marketing*. London: Routledge.

Kourtit, K., & Nijkamp, P. (2015) Cities of the future: Research challenges in the Urban Century. *Romanian Journal of Regional Science*, 9(1), 1–16.

Kudryavtsev,A., Stedman, R.C. and Krasny, M.E. (2012) Sense of place in environmental education. *Environmental Education Research*, 18(2), pp.229–250.

Malvika (2018) City branding. In *City Insight*, Planning Tank, available at: https://plannin gtank.com/city-insight/city-branding (accessed on 3rd December 2019).

142 *Sanjukta Sattar*

Mehta, S. (2006) *Maximum City: Bombay Lost & Found*. New Delhi: Penguin Books.

Middleton, A.C. (2011) City branding and inward investment. In: K. Dinnie (ed) *City Branding: Theory and Cases* (pp. 15–26). Basingstoke, Hampshire, UK: Palgrave Macmillan.

Maharashtra Tourism Development Corporation (2015) Ganesh festival. *Maharashtra Tourism*, available at: https://www.maharashtratourism.gov.in/maharashtra/festival/ganesh-festival (accessed on 30th November 2019).

Narayan, K. (2016) CityBiz: Mumbai has most foreign visitors. *The Indian Express*, September 24, 2016, available at: https://indianexpress.com/article/cities/mumbai/citybiz-mumbai-has-most-foreign-visitors-tourism-economy-3047098/ (accessed on 1st December 2019).

Parmenter, G. (2011) The city branding of Sydney. In: Keith Dinnie (ed) *City Branding: Thoery and Cases* (pp. 3–7). Basingstoke, Hampshire, UK: Palgrave Macmillan.

Prakash, P. (1993) The making of Bombay: Social, cultural and political dimensions. In *Economic and Political Weekly*, 28 (40), 2119–2121.

Punathambekar, A. (2013) *From Bombay to Bollywood: The Making of a Global Media Industry*. New York : New York University Press.

Riza M. (2015) Culture and city branding: Mega-events and iconic buildings as fragile means to brand the city. *Open Journal of Social Sciences*, 2015 (3), 269–274.

Riza, M., Doralti, N., & Fasli, M. (2012) City branding and identity. *Procedia: Social and Behavioral Science*, 35 (2012), 293–300.

Shaban, A. (2019) Mumbai. In A.M. Orum (ed) *The Wiley Blackwell Encyclopaedia of Urban and Regional Studies*. Hoboken, NJ: Wiley Blackwell. doi:10.1002/9781118568446.eurs0207

Statista Research Department (2018) Film industry in India: Statistics & facts, available at: https://www.statista.com/topics/2140/film-industry-in-india/ (accessed on 3rd December 2019).

Szondi, G. (2011) Branding Budapest. In Keith Dinnie (ed) *City Branding: Thoery and Cases* (pp. 3–7). Basingstoke, Hampshire, UK: Palgrave Macmillan.

Tata Mumbai Marathon (TMM) (n.d.) The possible dream, available at: https://tatamumbaimarathon.procam.in/about-event/event-history (accessed on 4th December 2019).

Uysal, U.E. (2013) Branding Istanbul: Representations of religion in promoting tourism. *Place Branding and Public Diplomacy*, 9(4), 1–13.

TATA Literature Live (2019) available at: https://2019.tatalitlive.in (accessed on 4th December 2019)

https://10times.com/mumbai-in/fashion-accessories (accessed on 4th December 2019)

https://timesofindia.indiatimes.com/life-style/fashion/shows/highlights-of-bombay-times-fashion-week-2019/articleshow/68547450.cms (accessed on 4th December 2019).

https://www.thehindubusinessline.com/news/variety/Lakmé-Fashion-Week-puts-the-spotlight-on-Indian-designers/article20112381.ece (accessed on 4th December 2019)
Information Classification: General

9 "It's coming home!"

Leveraging legacies in the City of Sails

Richard Keith Wright and Christopher Barron

Introduction

The drive to profit from the production of a positive brand identity has attracted significant academic and industry attention over the past couple of decades. Major Sports Events (MSEs), for example, are increasingly being viewed as a way to promote, position, and brand destinations (Dimanche, 2003). Since the turn of the century, however, there has been a significant increase in the amount of critical discourse and widespread doubt around the long-term return on investment gained from leveraging the legacies attached to hosting MSEs. Herstein and Berger (2013) suggest that the alignment of destination brands and communications with MSE must take a more holistic view, focusing on longer-term investment as a commercial foundation to gain longer lasting benefits. To date, much of the focus has been on the tangible, socio-economic, legacies attached to the arrival of high-yielding sport tourists. According to Weed (2008), sport tourism incorporates all forms of active and passive involvement in a sporting activity, participated in casually or in an organised way for non-commercial or business/commercial reasons, that necessitate the travel away from home and work locality.

The aim of this chapter is to showcase the long-established connections between the America's Cup and the city of Auckland, the city of sails, home to over a quarter of New Zealand's 4.5 million residents. The chapter was written prior to the emergence of the global COVID-19 pandemic. At the time of writing, two of the planned America's Cup World Series Regatta's had been cancelled and the fate of Auckland's 2020 Christmas regatta, the 2021 Prada Cup and the 36th America's Cup defence (AC36) remained largely unknown. According to an online news article published on Sail-World in March 2020, the co-ordinated financial investments made at a regional and national level could still "deliver a very significant and timely return" and the AC36 could play "an important part of the exit strategy for New Zealand to right a capsized and near-sunk tourism industry" (Gladwell, 2020).

The chapter starts with a review of the sports event literature, with a particular focus upon the event leveraging and legacy and destination branding. It then offers a case study that connects Auckland's emergence as an award-winning major sports event tourism destination to the world's oldest international sports trophy.

The conclusion and recommendations focus on the future potential attached to a nostalgia-driven revival of the City of Sails brand identity.

Auckland: The city of sails

In 2014, the city of Auckland, New Zealand, was listed in the Lonely Planet's "Best in Travel" guide as one of the world's Top 10 Cities (ATEED, 2014). It was also awarded third place at the annual SportBusiness Ultimate Sports Cities Awards, beaten on this occasion by London and Melbourne (The Aucklander, 2014). At the same awards ceremony, New Zealand's largest city was ranked first in a number of other categories, including the most ambitious city in the Asia Pacific region, the best medium-sized sporting city, the best event legacy, the best event security and the best homegrown event (The Aucklander, 2014). In 2012 and 2013 the City was placed in the top three sports event destinations in the world by the independent judges of the annual International Sports Event Management Awards, beaten only by London in 2012 (One News, 2013).

According the Chairman of Auckland Tourism, Events and Economic Development (ATEED), the regional council-controlled development agency, the awards and accolades picked up between 2012 and 2014 "demonstrate that Auckland is delivering on our ambitious Major Events Strategy ... our growing reputation as an events city is a boost to the visitor economy, which in turn is an enabler of social and cultural development of Auckland" (ATEED, 2014). Auckland's *City of Sails* slogan was first conceived in 1982 for a newly opened city centre Sheraton Hotel (Smith, 2019). In 1985, Campbell Advertising, with agreement from the Hotel, gifted the concept to the Auckland Council. The ACC promptly handed it over to Tourism Auckland, the regional tourism organisation that was swallowed up by the creation of ATEED, who used it extensively during the nation's early involvement with the America's Cup (Smith, 2019). Drinnan (2016) reported that the distinctive destination marketing campaign was discontinued around 2008 as a result of it being deemed too narrow in focus and unreflective of modern, multicultural Auckland.

In 2016, ATEED's head of destination marketing further justified the need for a new brand image/identity, claiming that "If Auckland is to compete in the global marketplace to attract high-value visitors, talented migrants and innovators, major sporting and business events, international students, multinational businesses, it must have a unique global brand"(Carroll cited in Drinnan, 2016). At the same time, however, ATEED's senior leadership team were forced to publicly defend spending $500,000 on the creation of a new brand narrative titled "Tamaki Makaurau, Auckland, The Place Desired by Many". On 26 March 2018, a Host City Appointment Agreement (HCAA) was signed between America's Cup Events Ltd (ACE), Emirates Team New Zealand (ETNZ), Ministry of Business of Innovation and Employment (MBIE) and Auckland Council confirming Auckland as the host of the 36th America's Cup, scheduled for March 2021 (ATEED, 2019b). According to Auckland Tourism, Events and Economic Development (ATEED), the city will also deliver a series of preliminary

America's Cup World Series Regattas, an America's Cup Christmas Regatta and the PRADA Cup Challenger Selection Series (ATEED, 2019a). The very recent and dramatic impacts associated with the global COVID-19 pandemic arose after completing acquisition of research data, so such impacts are not accounted for in this publication. Nevertheless, the consequences for major sports events are significant and are likely to have a lasting impact. While this presents both challenge and opportunity for New Zealand, the author is of the view that opportunities far outweigh the challenges, so, directing effort towards those opportunities presents an opportunity for brand-building and to stimulate the heavily impacted international tourism markets for Auckland and New Zealand.

Sports event marketing: Leveraging of Legacies (LoL)

The regular hosting of major sporting events (MSEs) has helped to shape the profile and popularity of many cities over the past century, if not longer (Getz, 1997; Smith, 2012). Chalip and Costa (2005) conclude that a growing desire to attract the (sport) event tourist has grown significantly over the past 20 years with many local, regional, and national tourism organisations viewing major events as a powerful vehicle for attracting new investment, new infrastructure and new visitors to a region. This observation was echoed by Lee and Kim (2014) who suggest that sports tourism can be leveraged from the well-established reputations of MSE and that new or existing destination brand identities can be revealed, reshaped, or revitalised through strategic event-related legacy development (Lee and Kim, 2014).

The terms "leverage" and "legacy" have become commonplace within the public and professional place event marketing discourse over the past decade, with obvious partnerships created through unique events and host cities to develop social, economic, and image/branding legacies. The rise of place marketing was discussed by Cudny (2019), as neoliberal approach being well recognised as "popular in urban development strategies implemented in the western world in recent decades". Cudny (2019, p.16) defines place marketing as "activities aimed at enriching the product of a given city by offering its consumers a well-chosen, interesting and diverse portfolio of events". Within the scope of this definition, they include event-related promotions, especially that which targets the recipients of urban products and portrays a positive image of the host destination. In this instance, event marketing is further defined as "a comprehensive group of activities related to city branding and city promotion" (Cudny, 2019, p.17). These marketing activities can be part of a longer-term urban development strategy that incorporates planning, organising, and promoting urban events. The outcome, according to Cudny (2019), should be the realisation of broadly defined host city development goals that service the various needs of the consumer.

Through the AC36 Auckland has several opportunities to leverage the nostalgia attached to America's Cup event to reinforce the city's brand through targeted place marketing, stimulate urban re-development, and nurture domestic and foreign investment. There is a clear opportunity through global media exposure attached the AC36 for Auckland and New Zealand to use the AC36 as a platform to reinforce/manipulate the city's identity. By simultaneously targeting the international audience and drawing on the rich historical and cultural maritime connections between the city and its waters to heighten destination profile, the event can be leveraged to increase brand awareness, identity, and reputation, while drawing on the unique association between Auckland and the America's Cup. This awareness (past and present) continues to be a powerful influence on international perceptions of Auckland City and on the City's own identity.

Over the past 30 years, every Summer and Winter Olympic Games, FIFA World Cup and Commonwealth Games have provided academics operating within the sport, event and tourism fields the chance to observe, access, evaluate, and articulate the expectations and experiences of local residents, spectators, tourists, sponsors, volunteers, and a range of other key sports event stakeholders. Despite a myriad case studies documenting both good and bad leveraging of legacy (LoL), however, one doesn't have to look too hard, or for too long, to uncover fresh evidence of destination marketers/managers failing to deliver on their pre-event promises and failing to maximise the potential long-term benefits attached to hosting MSEs. The common acceptance that legacies are an unavoidable consequence of staging all/any events has led to an increase in the number of articles focused on ways in which outcomes/opportunities can be strategically manufactured, manipulated, and maintained (i.e. leveraged), thus extending the finite length/life of a "once-in-a-lifetime" event to ensure the local stakeholders see a noticeable, equally note-worthy, return on their investment.

Unlike the over-utilised notion of "legacy", the strategic "leveraging" of major sports events is seen as being a more recent "phenomenon", with empirical research described as being "sparse and limited to mega and hallmark events in large cities" (O'Brien, 2007, 141). Chalip (2004, 228) defines event leveraging activities as those "which need to be undertaken around the event itself" and those "which seek to maximize the long-term benefits from events". Preuss (2007, 211) defines an event legacy as "all planned and unplanned, positive and negative, tangible and intangible structures created for and by an event that remain longer than the event itself". For cities and countries hosting MSEs, the focus of legacy-leveraging continues to be an integral part of event planning (Grix, 2014). The legacy planning process requires a wide-ranging investigation of the political, economic, social, technological, and environmental opportunities (Knott et al., 2016) associated with the hosting of MSEs. Grix (2014) suggests that legacy is a more dynamic approach in leveraging MSEs through a shift in planning towards the inclusion of assessing both shorter- and longer-term impacts of an MSE on the host destination. This more robust approach entails an impact assessment of MSEs to account for legacy implementation prior to, during and after the event to generate specific outcomes (Ritchie and Adair, 2006). Thus, legacy planning has resulted in a focus shift from short-term impacts such as job creation and attraction of spectators to a focus on the events ability to capitalise on short- and long-term opportunities associated with the event (Gripsrud et al., 2010).

Pike (2005) suggests that having well-grounded destination branding strategies to leverage MSE are central in establishing an effective brand. Similarly, Lantos

(2005) argued for the incorporation of brand equity, identity, positioning, personality, character, culture, and image within long-term planning, thus enabling the generation of more positive consumer experiences. The incorporation of a well-grounded purpose/vision was also seen a pivotal to aligning external and internal stakeholders to brand heritage, culture, people, and values, philosophy (Lantos, 2005). Matheson and Victor (2006) support this, suggesting that a destination brand must connect to a wide range of target markets and successfully differentiate the destination brand from its competitors. Building upon this idea further, Hankinson (2007) proposes that the use of corporate branding strategies have become more prevalent in co-branding to influence destination image. He also notes how corporate branding principles offer a fundamental framework for practitioners to consider in developing co-production of the place product. These principles include the co-production of the place product, the co-consumption of the place product, the variability of the place product, and the legal definition of place boundaries (Hankinson, 2007). More recently, Donner and Fort (2018) concluded that it is the multi-layered foundation and formulation of values that form the pillars of legacy leveraging, and destination personality. The following section looks at the foundations and formulation of Auckland's destination personality.

ATEED and the America's Cup (the Auld Mug)

Auckland Tourism, Events and Economic Development (ATEED) was established in November 2010. At the launch of the region's first Major Event Strategy (MES) in 2012, its Chairman predicted that Auckland was "poised to become a global events destination" and that the MES offered a blueprint to help achieve this ambition (ATEED, 2012). Likewise, at the same event, Auckland's Mayor referred to the strategic securement of major events as being "pivotal" in terms of helping the council "transform Auckland into the world's most liveable city" by the year 2040 (ATEED, 2012).

Between 2011 and 2013 Auckland's Major Events Strategy (MES) is said to have stimulated a 35% increase in GDP impact, growing from a $28.9 million return in 2011/12 to $39.1 million in 2012/13 (ATEED, 2014). Furthermore, 514,000 visitor nights were generated solely by the city's events portfolio over the same period (ATEED, 2014). In 2014, the city's General Manager of Destination and Marketing told reporters that the international recognition reinforced Auckland's position as "a true global city" (ATEED, 2014). A year later, ATEED's Chief Executive claimed that "more people are visiting Auckland than ever before and they're generally spending and doing more".

The hosting of major sports events such as the 2012 World Triathlon Grand Final, 2015 Cricket World Cup and 2015 FIFA Under 20s World Cup were credited for pumping more than $85 million into the regional economy, off an investment of $14 million (The Aucklander, 2014). In 2015, having established Auckland as a global city, ATEED also embarked on their own journey of discovery, a research campaign aimed at capturing an authentic and distinctive Auckland story. The person employed to manage the transition from Global Auckland to the

Auckland Story stressed the importance of having an underlying story and noted the successful destination marketing campaigns launched in London, New York, and Brisbane. She concluded, however, that;

> Most people in the world have barely heard of New Zealand, let alone Auckland. And then their response usually is 'oh I'd love to come, but it's so far.
>
> (Bridgwater cited in Idealog, 2015)

The "Place Desired by Many" brand was said to have been inspired by a translation of Auckland's Maori name, Tamaki Makaurau, although Wilson (2016) later suggested that this translation wasn't fully accurate. The rebrand was labelled a waste of money by a number of local councillors and never endorsed or approved by the city's mayor (Orsman, 2016; Radio New Zealand, 2016). Wilson (2016) defended the costs, however, and compared the $500,000 invested by ATEED "Auckland Story" over a two-year period with that spent by Tourism New Zealand on their own "New Zealand Story" campaign (NZD$2 million). He accepted that the slogan they had chosen wasn't particularly memorable or distinctive, but also reminded his readers that it was ATEED's job "to sell the city to the world".

There is no shortage of literature on the America's Cup, or it's connection to Auckland, New Zealand, the only city to outside of North America and Europe to host it on more than one occasion (Sefton and Keating, 2017). Steve Watters (2018) notes the popularity of recreational and competitive sailing in New Zealand, claiming sailors from the sparsely populated and geographically isolated South Pacific nation have "forged a formidable reputation on the water, from the Olympics to ocean classics such as the One Ton Cup, various iterations of round-the-world races and the America's Cup". He also cites how, unlike America, Sailing is big business for New Zealand-based boatbuilders and designers with the export of boats, marine technology, and equipment said to be worth $850 million to the economy in 2008 (Watters, 2018). Watters (2018) offers an extensive history of New Zealand's involvement in the America's Cup, an open water regatta frequently referred to as being the pinnacle of sailing and oldest active trophy in international sport (Fisher and Livingson, 2013; Hamilton, 2013; Rayner, 2003; Sefton and Keating, 2017; Simpson, 2012).

Despite its long history, Watters (2018) concludes that "many see the America's Cup as a trophy for the wealthy, with one commentator describing it as 'a pissing contest between the world's richest men'". The America's Cup, also known as the Auld Mug, is named after a boat and not the country from where it came (Simpson, 2012). The legacy of the international regatta began in front of Queen Victoria on the Isle of Wight in the August of 1851 (Raynor, 2003). An American commodore, John Cox Stevens, sailed a Schooner called America to victory against a fleet of 14 British built boats from the Royal Yacht Squadron and returned to Newport, Rhode Island with a silver claret jug known at the time as the 100 Guinea Cup (Simpson, 2012; Thompson and Lawson, 1986). Hamilton (2013) cites that the commodore was going to melt down the trophy to make medallions for his sailors but later decided to donate it under a Deed of Gift, stating that it was to be a perpetual challenge cup for friendly competition between nations (Thompson and Lawson, 1986).

The first defence of the original claret jug took place in 1857 and between 1870 and 1983 the New York Yacht Club successfully retained it on 24 further occasions (the longest winning streak in sport) (Fisher and Livingston, 2013; Simpson, 2012). There were six challenges prior to the turn of the twentieth century, four originating from the British Isles and two from Canada (Thompson and Lawson, 1986). The current America's Cup trophy was made by the royal jewellers Gerrard & Co in London in 1948 and the additions of two pedestals in 1958 and 1992 means that it is currently 1.1-metre-high, weighing 14kg (Bunting, 2017). In 1970, a separate event and trophy was required to determine who would be the next challenger. In 1983, the Challenger Selection Series was rebranded as the Louis Vuitton Cup and, having earned the right to race the New York Yacht Club, a crew representing the Royal Perth Yacht Club became the first sailors to leave America with the America's Cup in their possession (Simpson, 2012).

In 1987, the 26th America's Cup was hosted in Freemantle, Western Australia, and for the first time featured a challenger from New Zealand, a nation with more sailboats per capita than any other (Smith, 2002; John and Jackson, 2011). The New Zealand team that competed in the 1987 Louis Vuitton Cup represented Whitianga's Mercury Bay Boating Club (Daly, 2013). It was one of 13 syndicates from 6 countries (the others travelled from Canada, France Italy, the U.K. and the U.S.) and was primarily funded by an merchant banker named (Sir) Michael Fay (The Informer, 2017). Belgian-born Sydney-based businessman Marcel Fachler is reported to have paid the $16,000 entry fee (Daly, 2013). New Zealand's first attempt to challenge for the America's Cup came to an end in the Louis Vuitton Cup Final. The crew of New Zealanders representing a small town sailing club with less than 100 members were beaten by one of the six American syndicates sent across the Pacific Ocean to reclaim the Auld Mug (Daly, 2013).

The Mercury Bay Boating Club's second attempt to win the America's Cup took place in the September of 1988, following a legal challenge that resulted in a "deed of gift" challenge (Fare et al., 2003; Sefton and Keating, 2017). Two races were run between the defenders (the San Diego Yacht Club) and the challenger of record (Team New Zealand). The American team exploited a loophole in the rules and opted to enter a much faster catamaran against the larger mono-hulled New Zealand boat (Sefton and Keating, 2017). Although the defenders won both races, the outcome was challenged and on 28 March 1989 the New York State Supreme Court disqualified the San Diego Yacht Club, awarding the Auld Mug to the New Zealand team (Daly, 2013; Fare et al., 2003). Within 12 months of this verdict, however, the San Diego Yacht Club were reinstated as winners and the trophy was sent back to California (Sefton and Keating, 2017).

In 1992, the Mercury Bay Boat Club made a third and final attempt to win the Auld Mug, only to be beaten by the Italian syndicate in the final of Louis Vuitton Cup (Simpson, 2012). Three years later two New Zealand syndicates entered the America's Cup for the first time, one of which represented the Tutukaka South

Pacific Yacht Club. The other represented Auckland's Royal New Zealand Yacht Squadron. The 1995 Royal New Zealand Yacht Squadron's challenge was led by 1989–1990 Whitbread Round the World Race winner (Sir) Peter Blake and Skippered by 1984 Olympic Gold Medallist (Sir) Russell Coutts. Their boat, Black Magic, successfully obtained the Louis Vuitton Cup before winning the America's Cup for the first time without losing a race (Gladwell, 2017). In March 1997, the trophy was attacked with a sledgehammer while on display at the Royal New Zealand Yacht Squadron clubhouse (Daly, 2013). As a consequence, according to Bunting (2017), the repaired trophy "is always guarded while on display… usually flanked by two white-gloved security men". It also has its own special Louis Vuitton trunk, presented on the trophy's 150th birthday, and travels business or first class (Bunting, 2017).

Orams and Brons (1999) evaluated the potential impacts of New Zealand's biggest city hosting the 2000 America's Cup defence, calling it the biggest sports event to be hosted in the country since the 1990 Commonwealth Games (Orams and Brons, 1999). Their conclusion noted "the immense potential which exists for the nation to both capitalise on the event, and to upgrade the existing infrastructure and facilities for residents and tourists alike" (1999, 23). They also suggested that the event could provide "a great stimulus for visitors to travel to the country, and may help to promote New Zealand as an attractive tourist destination in the long term" (p.23). Furthermore, they acknowledged how the arrival of the event had "facilitated the need for overdue rejuvenation and development of the Auckland waterfront region" (p. 23) and proposed that delivering an event of this size would enhance the country's chances of hosting more major sports events in the future.

The America's Cup website claims that

> In the decades leading up to the late 1990s, the downtown waterfront area had witnessed the development and demise of a range of commercial and industrial uses including timber milling, boat building and fishing… Along with a renewed Kiwi euphoria for everything that sails on the water - hosting the America's Cup became the catalyst for a rejuvenation of the rundown Viaduct Harbour area. In just a few years, the Viaduct was transformed from a dilapidated area that Auckland citizens mostly avoided to a buzzing social and cultural hub, referred to as the dynamic heart of the city.
>
> (America's Cup, 2019a)

Auckland's newly rejuvenated Viaduct Harbour, the focal point for off-the-water activity during both the 1999 Louis Vuitton challenger series and the 2000 America's Cup, has also played host to a number of major international sailing events held in Auckland over the past 20 years, including the America's Cup defence in 2003, the Volvo Ocean Race, the Millennium Cup, and Vendee Globe round-the-world race (America's Cup, 2019a).

In 2000, at the 30th America's Cup, Team New Zealand became the first non-American syndicate to successfully defend the Auld Mug, sailing Black Magic to another 5-0 win, this time against the Italian Luna Rosa syndicate on Auckland's

Hauraki Gulf. Luna Rosa had earned the right to challenge Team New Zealand for the America's Cup by finishing ahead of 10 rival challengers from 7 countries, all of whom competed in a challenger series that lasted 4 months (Sefton and Keating, 2017). Having successfully retained the trophy, the 31st edition of the America's Cup was also hosted in Auckland, giving the city the opportunity to re-use the infrastructure built ahead of the 2000 regatta (Green, 1999), to reduce, if not remove, the risks/costs identified during the post-event evaluations conducted in 2000 (Orams and Brons, 1999) and to the reap the many socio-economic rewards attached to hosting a high yielding sports tourism-generating event (Barker et al., 2001).

In 2001, having stepped down as leader of Team New Zealand at the conclusion of the successful 2000 America's Cup defence, Sir Peter Blake was tragically murdered while on an environmental voyage up the Amazon River (Watters, 2018). During the same year, Barker et al. (2001) published a paper evaluating the economic impacts of hosting the 2000 regatta and offering further evidence of the long-term potential attached to New Zealand's retention of the Auld Mug for another three years. The paper shares the official pre-regatta regional and national forecasts of NZ$500–800 million for Auckland and NZ$1.3billion for New Zealand's economy and the amount of public money invested by Auckland Council (NZ$120 million) and New Zealand's national government (NZ12.5 million) (Barker et al., 2001). The authors propose that the hosting of special events, both sporting and non-sporting, represented a major component of Tourism Auckland's strategy to diversify and extend the length of stay of its visitor markets. They also acknowledge how the 2000 America's Cup was deliberately used to promote this cause to an international audience, "including markets that were difficult to target or unknown before the Cup" (Barker et al., 2001, p.90). The article concludes;

> The publicity generated by the Cup [AC30] and the associated exposure of New Zealand has undoubtedly created numerous tourism, trade, and investment opportunities. The recognition of the opportunities beyond the event is crucial to maximizing the latent potential of the event and its legacy into the future. Despite the financial shortfall of the Cup Village, indications suggest that the net success of the America's Cup in 2000 was beneficial to Auckland and that the true success of the event can only be understood in the longer term… As destinations such as New Zealand seek to utilize special events as a tourism development strategy, there are clear implications to be learned for the hosting of future events, which in 2003 will include the America's Cup… The success of the 2000 and 2003 America's Cup events will greatly affect Auckland's ability and willingness to host international events in the future.
>
> (Barker et al., 2001, p.90)

In 2003, the second Louis Vuitton Challenger series to be sailed on Auckland's Hauraki Gulf featured 120 races over a five-month period, involving 9 syndicates from 6 countries. It was won by the newly established Team Alinghi, created by Swiss billionaire Ernesto Bertarelli and represented the Société Nautique de Genève (SNG). At the helm of the Swiss boat was the former Team New Zealand skipper and two-time America's Cup winner Russell Coutts (Boshier, 2003; Rayner, 2003). Team Alinghi's tactician was former Team New Zealand employee Brad Butterworth and, in total, the crew of 18 that ended Team New Zealand's eight-year reign as the holders of the America's Cup contained 6 New Zealanders, all of whom had participated in the 1995 challenge and the 2000 defence (Boshier, 2003). Bertarelli's multi-national team also included two Americans involved in the successful San Diego defences of 1988 and 1992 plus several Olympians from Canada, Germany, Holland, and Italy (Raynor, 2003).

John and Jackson's (2011) article on the globalisation and corporate national- ism evident within the America's Cup reveals that "segments of the New Zealand public, perhaps influenced by the media, viewed the actions of these sportsmen as being selfish, akin to being a traitor". They add that critics "argued that these yachtsmen had betrayed 'the nation' which had contributed, both financially, and emotionally, to their personal success" and note that a group called 'Blackheart' was established to publicly shame the privately owned America's Cup syndi- cates who had benefited the most from defecting New Zealand sailors (John and Jackson, 2011). The Blackheart campaign, which primarily consisted of a couple of billboards and a website was closed down before the America's Cup began, however, after "some of the so-called 'traitors' were threatened with grievous bodily harm" (Rayner, 2003, 96, cited in Johns and Jackson, 2011).

The fact that a syndicate from Switzerland departed Auckland International Airport in 2003 with the Auld Mug meant that the next chapter of America's Cup long history was built around the staging of the first ever European-based defence. It also ensured the defenders would also be required to try and retain it in foreign waters (Switzerland being nation with no coastline). There were a number of changes to the racing protocols following SNG's victory in Auckland, with the most notable being the removal of the rules restricting the number of foreign sailors on each boat. This allowed the team owners with the deepest pock- ets to approach and employ the best people regardless of their nationality (Perry, 2004; Gladwell, 2017). The rules around the transfer of technology from prior syndicates were also modified, allowing new teams access to old information (Simpson, 2012). Finally, a new organising authority called AC Management was created and charged with the task of overseeing all aspects of the 32nd America's Cup, including the Challenger Selection Series (Sefton and Keating, 2017). The host city chosen by SNG was Valencia, Spain, and the new rules resulted in even more New Zealand sailors and boat builders being employed by teams based all over the world.

In 2007, ETNZ won the Louis Vuitton Cup for a second time, setting up an America's Cup rematch against Team Alinghi (Johns and Jackson, 2011). Once again, however, the Swiss boat proved to be too fast and the challengers returned to Auckland and a barrage of media questions regarding their future (Gladwell, 2017). The 33rd America's Cup was also hosted in Valencia, but – as with the 1988 Deed of Gift contest – there was no Louis Vuitton Cup and only two races between the defender and challenger. It was also an event overshadowed by litigations and multiple court hearings (Sefton and Keating, 2017). The 2010 Deed of Gift event was dubbed the battle of the billionaires (Daly, 2013) and Oracle's 2-0 victory gave the Russell Coutts a fourth successful America's Cup campaign. The challengers representing the Golden Gate Yacht Club featured three New Zealanders whilst the Swiss boat had six onboard, including Brad Butterworth who had taken over as skipper following Coutts' departure in 2007 (Simpson, 2012).

Coutts' fifth America's Cup victory occurred three years later in San Francisco when an Oracle Team USA Racing crew, featuring 3 New Zealanders, successfully defended the Auld Mug against the Louis Vuitton Cup winners ETNZ (a crew containing 11 New Zealanders). The New Zealand government are reported to have committed $36 million to the unsuccessful campaign (Watters, 2018). The challengers were 8-1 up in the final, needing only one more win (Gladwell, 2017). Oracle Racing, however, produced one of the biggest comebacks in sporting history to win the final 8 races, ensuring that the trophy remained the property of the Golden Gate Yacht Club for at least another 4 years. ETNZ returned to Auckland empty handed, but having won some new supporters. Some former sailors and local media commentators were critical of the amount of public money invested into the unsuccessful campaigns of 2007 and 2013, but others were buoyed by the extent to which a team of New Zealanders was still able to challenge the multinational team owned and resourced by a couple of billionaires (Gladwell, 2017; Watters, 2018).

In 2017, the 35th America's Cup took place in Bermuda, a British territory in the North Atlantic Ocean with a permanent resident population of around 65,000 (Stuff, 2014). New Zealand media reported that Bermuda beat San Diego for the hosting rights, having put together a US$77 million (NZ$98m) package for the event owners/organisers (Stuff, 2014). Within the same article, Oracle CEO Sir Russell Coutts claimed that it was the only bid in a time zone that satisfied the needs of the sponsors that could also offer a centralised base to house all of the competing teams (Stuff, 2014). Coutts also explained Oracle's decision to take the cup to a neutral venue as opposed to returning to San Francisco. He told reporters that the America's Cup is

> an international event, noting how; it's got international teams, and those teams have sponsors, and the broadcasters are an important part of that sponsorship. So we looked at some of those things and weighted them very, very highly... I'm really happy with where we are; this is a fantastic decision, and I'm absolutely convinced this is going to be a fantastic America's Cup. It's not a PR sell or anything like that; I believe it, I really do. It's going to be the best one yet.
>
> (Coutts cited in Stuff, 2014)

The event protocols changed on multiple occasions during the build-up to the 2017 America's Cup, which resulted in the withdrawal of two challengers. The new rules allowed the defenders to compete in the Louis Vuitton Cup series for the first time, ensuring that they got to test and compare their boat against those competing for the right to challenge them for the America's Cup (Gladwell, 2017). Unlike the other teams in Bermuda, however, the New Zealanders turned up on the eve of the event with a boat that was powered by their sailor's legs as opposed to their arms, replacing grinders for cyclors (Gladwell, 2017). ETNZ, sailing a boat called Aotearoa, won the America's Cup with a convincing 7-1 win and on 6 July 2017, paraded the America's Cup around the streets and waterfront of Auckland for the first time in seventeen years.

It's coming home: the Auld Mug is Auckland's Cup

In 1995, when the Auld Mug arrived in Auckland for the first time, Daly (2013) claims that "a throng estimated at more than 300,000 crammed into the downtown area for a victory parade. [Peter] Blake, who received a knighthood that year, thanked them for "this most stupendous, fantastic, terrific, marvellous New Zealand welcome". According to 2017 media reports, the most recent victory parade attracted "tens of thousands of joyful New Zealanders" (Roy, 2020). Newshub (2014) reported the number to be an estimated 80,000, excluding those who took their boats out onto the city's Waitemata harbour to watch the team. Within hours of the victory, questions were being asked as to where and when Team New Zealand would choose to defend the trophy, with many assuming that the event would be returning to Auckland, the city of sails, for a third time.

Winfield and Durhager (2012) suggest that the hosting opportunities afforded by the 2017 America's Cup regatta bolstered Bermuda's global stature and brand as an up-market destination and international sport event venue. Through the hosting of such sport events, destinations are given significant global broadcast audiences and media attention before, during, and after an event to communicate the nostalgia associated with the hosting of an event (Winfield and Durhager, 2012). Although Auckland's growing status as an award-winning sports event destination has reduced the impact of the city's geographic isolation, it has also increased the domestic divide between the city and the rest of the country. The arrival of AC36 therefore comes at a time where the long-term costs and benefits of Auckland's first ever major event strategy were being re-evaluated, providing a significant opportunity for Auckland's public-funded, event tourism economic development agency to further strengthen local awareness, attraction, and attachment to an ageing brand identity, whilst also providing a suitable justification for their existence and their ongoing pursuit of other tourism-generating major sports event.

Theunissen (2017) discussed the findings of an official report from New Zealand Ministry of Business, Innovation and Employment (MBIE) that referred to the America's Cup as "an iconic event in New Zealand's sporting history". The pre-event impact report estimated that every $1 invested in hosting the 2021 regatta will come back more than seven-fold by 2055 and between $600 million and $1 billion would be injected New Zealand's economy between

2018 and 2021 (Theunissen, 2017). Theunissen (2017) points out that the numbers quoted in the 2017 MBIE report outweigh the economic benefits attached to hosting the previous two America's Cups ($495m in 2000 and $529m in 2003) and the costs of hosting the 2021 event, estimated at being around $200 million (Theunissen, 2017). He also suggests that MBIE had to rely on assumptions primarily informed by the last Cup hosted in Auckland in 2003 to reach its estimates, examining the spending habits of the primary "expenditure groups", notably "the yachting syndicates themselves, superyachts, other visiting boats, international visitors and media" (Theunissen, 2017). The MBIE report concludes that

> Successive governments have seen the benefits that flow from investing in both the event itself [when held in New Zealand] and from investing in Team New Zealand ... even when the event is not going to be held in New Zealand. The flow-on effects for New Zealand's marine industry and "Brand New Zealand" are significant.

As was the case in 2000 and 2003, the 36th America's Cup is seen as an important opportunity for Auckland to shape tourism strategies and strengthen both City and National brands. Hosting the 36th America's Cup is also seen by both central and local governments as an opportunity to continue the wider downtown waterfront's transformation, initiated by hosting the 30th and 31st America's Cup 20 years earlier and a collaborative work programme was established in 2018 to prepare for and manage all aspects of the event (America's Cup, 2019b). The common vision of the stakeholders is for the 36th America's Cup to be

> an inclusive event, with a waterfront that will allow an experience accessible for everyone, connecting people to boats, bases and events, in a linear village that will spread across the waterfront from the Eastern Viaduct, to North Wharf and Wynyard Point.
>
> (America's Cup, 2019b)

Part of the vision, however, looks far beyond 2021 with the creation of new open spaces for people overlooking the harbour, opening up Wynyard Point to the public and leading the way for the future redevelopment of the regional destination park which will start in 2022 onwards. The permanent land and water spaces will create a legacy for existing and future water-based events making it easier for Auckland to bid for large international events (America's Cup, 2019b).

In 2018, the Ministry of Internal Affairs created a lottery fund specifically focused on helping New Zealand communities' benefit from the arrival of the 36th America's Cup. According to the Internal Affairs Minister Tracey Martin,

> The America's Cup is a global event, and having it here is a significant opportunity – one that can benefit towns and cities across New Zealand in the lead

up to, during and post-2021. While Auckland will be hosting the event, the Fund aims to spread community gains across the country.

(Ministry of Internal Affairs, 2018)

ATEED (2019c) believes that hosting AC36 in Auckland will generate significant and widespread benefits for Auckland (between $485 and $858 million derived from teams, sponsors, media, and visitors) and all of New Zealand (up to $1 billion), as well as an employment boost of between 4,700 and 8,300 jobs. Their programme partners are aiming to ensure that the event "accelerates the sustainable transformation of our communities, our water and our land" and "creates shared economic and wider benefits through connection, innovation and trade". They are also working to ensure that "every New Zealander has the opportunity to participate in and celebrate the America's Cup" and that "the rich cultural and voyaging stories of Tāmaki Makaurau and Aotearoa are shared and valued" (ATEED, 2019c).

"History is indeed repeating itself", according to one article located on the 36th America's Cup's official website (America's Cup, 2019b). The article documents the significant transformation that occurred to Auckland's "dilapidated fishing port" (the Viaduct Harbour) ahead of the 2000 regatta before claiming that the same thing was now happening to the adjacent Wynward Harbour area. In addition to delivering "up to an estimated $1 billion in value" (to the national economy), those responsible for selling the AC36 to locals and visitors also state that the major event

> provides opportunities for the city and the whole of New Zealand well beyond what happens on the water. America's Cup 36 (AC36) will be an opportunity to ignite Kiwis' passion for this great event and showcase our city and country, celebrating all that makes Auckland and New Zealand unique.
>
> (America's Cup, 2019b)

Four teams have committed to compete for the America's Cup in Auckland in 2021, with the Challenger of record being the Italian Luna Ross syndicate, representing the Circolo della Vela Sicilia. For the first time in America's Cup history, the challenger series will see a syndicate from the New York Yacht Club (American Magic, USA) racing against at syndicate from the Royal Yacht Squadron Racing (INEOS TEAM UK), whilst the fourth challenger is the Stars & Stripes Team USA, representing the Long Beach Yacht Club (ATEED, 2019d). For the first time this century, Russell Coutts will not be actively involved in any of the teams and there will be no challenger from the Golden Gate Yacht Club. Two additional challengers were publicly linked to the event, one from Malta and the other from the Netherlands, but neither was able to meet the entry requirements ahead of the deadline set by the defender and challenger of record (Luna Rosa) and failed to eventuate. Unlike the 2017 event, the defender will not compete in the challenger series.

Conclusion and recommendations: leveraging the legacies of the City of Sails

Co-branding between destinations and Major Sporting Events (MSEs) is well recognised as an important tool for destinations to leverage off, promote, position, and brand themselves (Hankinson 2007). So, at a time where Auckland's brand image and identity remains in question, the upcoming 2021 America's Cup (AC36) can be an opportunity to re-kindle a nostalgia-driven revival of the Auckland's "City of Sails" brand identity first launched in 1985. Along with this renewal of Auckland's identity, AC36 presents further opportunity to leverage the nostalgia attached to the world's oldest ongoing international sports trophy, to stimulate urban re-development, nurture domestic, and foreign investment, and to further develop through targeted positioning.

To conclude, Auckland, New Zealand, will soon join New York as being the only city to have hosted an America's Cup regatta on more than two occasions. This unique association has been and continues to be a powerful influence on international perceptions, and on the City's own brand identity. At the start of the 21st century, the exploitation of opportunities provided by successive America's Cup series served well to reinforce Auckland's brand positioning as the "City of Sails" and as a world class sport tourist destination (Johns and Jackson, 2011). In a practical way, the city has also been able to leverage these opportunities to revive and enhance the city waterfront (Green 1999; Orams and Brons, 1999). In the search for destination brand differentiation and a multi-cultural narrative for Auckland, the City has potentially underestimated the power and importance of the "City of Sails" brand.

The opportunity again presents itself for Auckland to exploit the sailing legacy of two past America's Cup regattas, both of which played a key role in providing opportunity for New Zealand's emerging sailing talent, and shaping the country's reputation as a leading sailing nation. When looking at the threat of the COVID-19 lockdown on the AC36, Gladwell (2020) notes how, in both 2000 and 2003, "the kiwi tourism authorities rather cheekily took advantage of the two events to fund/sponsor international media to experience tourist attractions and get the message back to their readers and viewers" (Gladwell, 2020). Regardless of the number of international sport event tourists permitted to enter New Zealand in 2021, or even the number domestic tourists able to travel from elsewhere in the country, the tangible and long-term legacy of the AC36 can already been seen through the improvement of Auckland's waterfront infrastructure and the number of new shops, restaurants and hotels opening up within walking distance of the shoreline. Furthermore, assuming that the event takes place, the images captured by the world's media will surely generate interest amongst locals and future tourists alike. Assuming that the locals choose to visit the spectator hubs and spectator viewing areas, AC36 should also further strengthen the city's existing event volunteer network ahead of Auckland's hosting of 2021 Rugby World Cup, 2022 Women's Cricket World Cup and 2023 FIFA Women's World Cup fixtures.

The evidence presented within the chapter confirms that AC36 will draw a global audience of spectators (if not tourists) during the lead up, which began in 2017. Unlike some of the other MSEs to have been hosted in Auckland in the past decade, the local event leveraging opportunities directly attached to AC36 can exist for several years, with the event itself running for four months (from December 2020 to April 2021). This aligns with the notion of place marketing suggested by Cudny (2019), highlighting how "products' merits are communicated to the wider audience and branded with the use of events, including sports events, festivals, business events". Further, the notion of (Cudny 2019) reinforces the importance of creating a "memorable happening which attracts visitors and makes a great impression on them". Without creating this memorable happening, events run the risk of not fully embracing the promotional opportunities afforded by events, thus diminishing the potential place marketing, socio-economic and tourism opportunities afforded by unique events.

Regardless of whether the event is able to take place in March 2021 (as scheduled), the City of Auckland has an opportunity to leverage the place marketing potential of the AC36 to reinforce identity and draw on the rich historical and cultural maritime connections between the City and its waters. A refreshed "City of Sails" brand that builds on nostalgia while at the same time embracing culture and history provides a potent opportunity for Auckland to stake out its identity and position. Rather than create new marketing campaigns that may take years to reinforce, Auckland should use its maritime history as a cornerstone of future marketing campaigns to leverage the nostalgia of past, present, and future Americas Cup success.

References

America's Cup (2019a) Building the next America's Cup and its legacy. Available at: https://www.americascup.com/en/news/443_BUILDING-THE-NEXT-AMERICA-S -CUP-AND-ITS-LEGACY.

America's Cup (2019b). History. Available at: https://www.americascup.com/en/history

ATEED (2014). It's official: Auckland's one of the world's best cities. Available at http:// www.aucklandnz.com/love/its-official-aucklands-one-of-the-worlds-best-cities.

ATEED (2019a). 36th America's Cup. Available at: https://www.aucklandnz.com/akl 2021/36th-americas-cup

ATEED (2019b). Event delivery. Available at: https://www.aucklandnz.com/akl2021/ 36th-americas-cup/ac36-event-delivery.

ATEED (2019c). Everybody wins. Available at: https://www.aucklandnz.com/akl2021/ 36th-americas-cup/benefits.

Barker, M., Page, S. & Meyer, D. (2001). Evaluating the impact of the 2000 America's Cup on Auckland, New Zealand. *Event Management*, 7(2), 79–92.

Boshier, R (2003) Using globalisation technology for localisation: How Schnak-Net reconstructed the America's Cup. *Sociology of Sport Online*, 6(1). Available at: http:// physed.otago.ac.nz/sosol/v6i2/v6i2_2.html.

Bunting, E. (2017). Things you never knew about the historic 'Auld Mug' America's Cup trophy. Available at: https://www.yachtingworld.com/americas-cup/things-never-knew -auld-mug-americas-cup-trophy-107498.

Chalip, L. (2004). Beyond impact: A general model for host community event leverage. In: Ritchie, B.W., & Adair, D. eds., *Sport Tourism: Interrelationships, Impacts and Issues*, pp. 226–252. Clevedon, UK: Channel View.

Chalip, L., & Costa, C. (2005). Sport event tourism and the destination brand: Towards a general theory. *Sport in Society*, *8*(2), 218–237.

Cudny, W. (2019). The concept of place event marketing. In: Cudny W. ed., *Urban Events, Place Branding and Promotion*, pp. 1–24. London: Routledge.

Daly, M. (2013). History of New Zealand in the America's Cup. Available at: http://www.stuff.co.nz/sport/9180810/History-of-New-Zealand-in-the-Americas-Cup.

Dimanche, F. (2003). The role of sports events in destination marketing. In: Keller, P., & Bieger, T. eds., *Sport and Tourism*, pp. 303–311. Proceedings of the 53 rd AIEST congress, St Gallen, Switzerland: AIES.

Donner, M., & Fort, F. (2018). Stakeholder value-based place brand building. *Journal of Product & Brand Management*, *27*(7), 807–818.

Drinnan, J. (2016). What's wrong with city of sails? Available at: https://www.nzherald.co.nz/business/news/article.cfm?c_id=3&objectid=11750168.

Fare, J., Foster, M., Manasse, D., Peter, H. & Tompkins, D. (2003). *Arbitration in the America's Cup: The XXXI America's Cup Arbitration Panel and Its Decisions*. The Hague: Kluwer Law International.

Fisher, B., & Livingston, K. (2013). *Sailing on the Edge: America's Cup*. San Rafael, CA: Insight Edition.

Getz, D. (1997). Trends and issues in sport event tourism. *Tourism Recreation Research*, *22*(2), 61–62.

Gladwell, R. (2017). *Lone Wolf: How ETNZ Stunned the World*. Auckland: Upstart Press Limited.

Gladwell, R. (2020). America's Cup shake out inevitable after cancellations and shutdowns. Available at: https://www.sail-world.com/news/227708/Americas-Cup-reset-after-tumultuous-fortnight.

Green, A. (1999). America's Cup 2000: The impact of the America's Cup Regatta on the Viaduct Basin. *Journal of Geography*, 107, 10–19.

Gripsrud, G., Nes, E., & Olsson, U. (2010). Effects of hosting a major-sport event on country image. *Event Management*, *14*(3), 193–204.

Grix, J. (2014). *Leveraging Legacies from Sports Mega-Events*. Basingstoke: Palgrave Macmillan.

Hamilton, S.L. (2013). *Xtreme Races: America's Cup*. Minneapolis, MN: ABDO Publishing Company.

Hankinson, G. (2007). The management of destination brands: Five guiding principles based on recent developments in corporate branding theory. *Journal of Brand Management*, *14*(3), 240–254.

Herstein, R., & Berger, R. (2013). Much more than sports: sports events as stimuli for city re-branding. *Journal Of Business Strategy*, *34*(2), 38–44.

Idealog (2015). What's the story? ATEED on the hunt for a global brand for Auckland. Available at https://idealog.co.nz/design/2015/03/whats-auckland-story-ateeds-bridgwater-quest-aucklands-global-yarn.

John, A., & Jackson, S. (2011). Call me loyal: Globalization, corporate nationalism and the America's Cup. *International Review for the Sociology of Sport*, *46*(4), 399–417.

King, A. (2003). The taxpayer's cup runneth over: Public sponsorship of yachting. *EcoNZ@Otago*, *11*, 1–4.

Knott, B., Fyall, A., & Jones, I. (2016). Leveraging nation branding opportunities through sport mega-events. *International Journal Of Culture, Tourism And Hospitality Research*, *10*(1), 105–118.

Lantos, G. (2005). The new strategic brand management: creating and sustaining brand equity long term. *Choice Reviews Online*, *42*(9), 42–53.

Lee & Jun H.K. (2014). Effects of servicescape on perceived service quality, satisfaction and behavioral outcomes in public service facilities, *Journal of Asian Architecture and Building Engineering*, *13*(1), 125–131.

Matheson, V. (2006). Mega-Events: The effect of the world's biggest sporting events on local, regional, and national economies. Economics Department Working Papers. Paper 68. Available at: http://crossworks.holycross.edu/econ_working_papers/68.

Ministry of Internal Affairs (2018). America's Cup: $20million fund created to provide Cup. Available at: legacy https://www.sail-world.com/news/212605/Yacht-clubs-able-to-apply-to-new-USD20m-Am-Cup-Fund.

New Zealand Herald Online (2003) Olympics: Delegates fail to tackle Mallard over America's Cup money. Available at: http://www.nzhearld.co.nz/storyprint.cfm?storyID=3502247.

Newshub (2014). 80,000 people cheered on America's Cup parade. Available at: https://www.newshub.co.nz/home/new-zealand/2017/07/america-s-cup-victory-parade-live-updates.html.

O'Brien, D. (2007). Points of leverage: Maximizing host community benefit from a regional surfing festival. *European Sport Management Quarterly*, *7*(2), 141–165.

One News (2013). Auckland ranked third in list of world's top sporting cities. Available at: http://tvnz.co.nz/national-news/auckland-ranked-third-in-list-world-s-top-sporting-cities-5702291.

Orams, M.B. & Brons, A. (1999). Potential impacts of a major sport/tourism event: The America's Cup 2000, Auckland, New Zealand. *Visions in Leisure and Business*, 18 (1), 14–28.

Orsman, B. (2016). Auckland's new $500,000 brand not so desired. Available at: https://www.nzherald.co.nz/nz/news/article.cfm?c_id=1&objectid=11746553.

Perry, N. (2004). Boots, boats, and bytes: Novel technologies of representation, changing media organisation, and the globalisation of New Zealand sport. In: Horrocks, R., & Perry, N. eds., *Television in New Zealand: Programming the Nation*, pp. 291–301. Melbourne: Oxford University Press.

Pike, S. (2005). Tourism destination branding complexity. *Journal of Product & Brand Management*, *14*(4), 258–259.

Preuss, H. (2007). The conceptualisation and measurement of mega sport event legacies, *Journal of Sport & Tourism*, 12(3–4), 207–228.

Radio New Zealand (2016). Auckland slogan faces scrap heap. Available at: https://www.rnz.co.nz/news/national/317939/auckland-slogan-faces-scrap-heap.

Rayner, R (2003). *The Story of the America's Cup: 1851–2003*. Auckland, NZ: David Bateman.

Ritchie, B., & Adair, D. (2006). *Sport Tourism*. New Delhi: Viva Books.

Roy, E. (2020). America's Cup parade: New Zealand crowds brave thunderstorm to welcome crew. Available at: https://www.theguardian.com/world/2017/jul/06/americas-cup-parade-new-zealand-crowds-brave-thunderstorm-to-welcome-crew.

Sefton, A. & Keating, L. (2017). *Exposed: The Dark Side of the America's Cup*. London: Bloomsbury.

Simpson, R.V. (2012). *The Quest for America's Cup: Sailing to Victory*, Charleston, SC: The History Press.

Smith, A. (2012). Sporting a new image?: Sport-based regeneration strategies as a means of enhancing the image of the city tourist destination. *Routledge Online Studies on the Olympic and Paralympic Games*, *1*(45), 109–124.

Smith, N. (2002). Passing the cup: The meaning of the America's Cup at the global table. *Harvard International Review*, *24*(1): 30–33.

Smith, R. (2019). https://www.quora.com/Why-is-Auckland-called-the-city-of-sails.

Soutar, G.N. & Mcleod, P.B. (1993). Residents' perceptions on impact of the America's Cup, *Annals of Tourism Research*, *20*(3), 571–582.

Stuff, (2014). America's Cup wooed to Bermuda with a $98m package. Available at: https://www.stuff.co.nz/sport/other-sports/63969278/americas-cup-wooed-to-bermuda-with-a-98m-package.

The Aucklander (2014). Auckland third best sporting city in the world. Available at: http://www.nzherald.co.nz/aucklander/sport/news/article.cfm?c_id=1503376&objectid=11236459.

The Guardian (2017). America's Cup parade: New Zealand crowds brave thunderstorm to welcome crew. Available at: https://www.theguardian.com/world/2017/jul/06/americas-cup-parade-new-zealand-crowds-brave-thunderstorm-to-welcome-crew.

The Informer (2017). Mercury Bay's proud association with the America's Cup. Available at: https://www.theinformer.co.nz/feature/mercury-bays-proud-association-with-the-americas-cup

Theunissen, M. (2017). Does the cost of America's Cup tally up for Auckland? Available at: https://www.nzherald.co.nz/business/news/article.cfm?c_id=3&objectid=11946282.

Thompson, W .M. & Lawson, T.W. (1986). *The Lawson History of the America's Cup: A record of Fifty Years*, Southampton: Ashford Press Publishing.

Tourism New Zealand (2018). Auckland: Waterborne in the City of Sails. Available at: https://media.newzealand.com/en/story-ideas/auckland-waterborne-in-the-city-of-sails/.

Watters, S. (2018). New Zealand and the America's Cup. Available at: https://nzhistory.govt.nz/culture/americas-cup.

Weed, M. (2008). Sports tourism experiences. *Journal of Sport & Tourism*, *13*(1), 1–4.

Wilson, S. (2016) Calm Down, NZ Herald. The new Auckland slogan search was fine. Available at: https://thespinoff.co.nz/auckland/17-11-2016/the-slogan-desired-by-not-many-if-any-simon-wilson-on-aucklands-new-marketing-campaign/.

Winfield, M., & Durhager, P. (2012). *Americas Cup Bermuda Legacy Impact*. Bermuda. Available at: http://11thhourracing.org/wp-content/uploads/2017/12/35thamericascupbermudalegacyimpact.pdf.

10 Conclusions

Waldemar Cudny

This book is a second edited volume, devoted to the role of event marketing in urban spaces development and promotion, edited by Waldemar Cudny. Event marketing is presented in the scientific literature in various ways. On the one hand, it is defined as a procedure of successful selling of events. This approach includes ideas on how to create interesting events that will suit the market demand and attract crowds of spectators. It also characterises how events should be promoted (via traditional and social media) to sell well and attract more event-goers (see: Hoyle 2002). On the other hand, event marketing is often associated with the marketing of firms and products. In this case, event organisers focus on the creation of marketing messages and memorable happenings which will be presented to the potential customers and create a positive brand association with a certain product and firm during a well organised event (see: Kotler and Armstrong 2010; Jaworowicz and Jaworowicz 2016).

However, another understanding of event marketing exists in the literature and it refers to urban places i.e. cities and towns. In this case, event marketing is associated with place marketing, place branding, and urban promotion. Events are part of place branding plans and marketing strategies of cities and towns because they develop the urban product by offering entertainment and leisure opportunities to inhabitants and tourists (Degen and Garcia 2012). Events boost revenues for local firms (from the event and hospitality sector) and create employment (see: Smith 2012; Cudny 2016, 2019). Moreover, they influence the perception of a specified place i.e. create its brand. Events are often presented in the media both traditional and social media and are subject to the word-of-mouth communication. Thus they are a unique tool of urban promotion that helps selling places (see: Wilson 2004; Cudny 2016; Broudehoux 2017).

The issue of place event marketing concerning urban places was presented in a complex way in an edited volume devoted to this subject published by Cudny (2020b). The first chapter of the aforementioned book encompassed a detailed theoretical elaboration of what is place event marketing in urban spaces. The chapter was followed by further chapters with case studies from five continents (North and South America, Europe, Asia, and Africa). These case studies presented how events create interesting urban place products attracting tourists and giving host places amazing promotional opportunities and benefits. Eventually as stated by Cudny (2020a) place event marketing is defined as a twofold procedure encompassing

1. creation of interesting experience-based urban products that influence socio-economic urban development and create its image and brand,
2. promotion of cities and towns with the use of events organised in them.

This edited volume regarding place event marketing in the Asia Pacific Region is a second publication presenting the research issue of place event marketing edited by Waldemar Cudny. This book connects both of the elements of place event marketing presented by Cudny (2020a). It characterises how various events organised in selected cities across the Asia Pacific region influence the development of tourism and create other socio-economic legacies. Moreover, the book analyses how the organisation of events contributes to changes in the image of host places and their brand creation.

The idea behind this book was to supplement the previous publication (i.e. Cudny 2020b) with case studies from the region which is currently undergoing vast economic development i.e. Asia and the Pacific (see: Fang and Chang 2016). Moreover, the region of Asia and the Pacific is nowadays very important in terms of tourism and attracts growing crowds of visitors. The region in 2018 was presented by the UNWTO data as the second (according to the number of international arrivals) tourist region in the world (International Tourism Highlights. Edition 2019, 2019). The number and diversity of events organised there are growing significantly in recent years too (see: Tolkach et al. 2016). However, the amount of scientific analysis events and their role in urban marketing is rather short. Eventually, the will to fill in the research gap on place event marketing in regard to the Asia Pacific region stood behind the decision to write this book.

The main research aims of this volume included a presentation of how different kinds of events could be successfully used in place event marketing strategies encompassing promoting and branding of different types of urban places with the use of events. It also included a presentation of how public and non-governmental institutions can implement place event marketing activities in cities. Another research aim was the presentation of how different governmental and non-governmental organisations may support the event-marketing agenda. The book also aims to show the effects of place event marketing on urban place promotion and branding and presenting a variety of empirical case studies characterising it on the example of cities and towns from the Asia Pacific region.

This book focuses on eight selected areas from the Asia Pacific region. It encompasses the following countries and urban spaces – China (Hangzhou and Shanghai cities), Japan (Okinawa Prefecture and Saitama city), India (Kolkata and Mumbai cities), New Zealand (Auckland city), and selected examples of cities from the ASEAN group of countries (Association of Southeast Asian Nations). The events under study represented a great variety of types. The individual chapters analysed sports events, religious and cultural festivals, as well as MICE events.

In Chapter 2, Shue-Wei Tsai presented the analysis of the G20 summit in Hangzhou and its impacts on the brand of the city and China. The G20 is a group established in 1999, which consists of the richest and the most important countries in the world and a representative of the European Union. The aim of the group includes among others to carry out activities regarding the development of the global economy, sustainable growth, introduction of clean energy that cannot be implemented within individual countries. The aim of the annual G20 summits, attended by representatives of the members of this group, is to discuss a common economic policy in a global system.

Hangzhou city is located in eastern China in Zhejiang Province and is inhabited by 7.7 million people. The last decades brought a distinctive growth in Hangzhou's economy and the city became an important hub of industry, services, and university education. It also became the headquarter of Chinese e-commerce giant Alibaba Group. The city was less recognised internationally, and despite its size, it was not included in the group of most important Chinese cities (first-tier cities).

The G20 summit held in Hangzhou in 2016, was an important branding and promotional event leveraging the city and its brand internationally and domestically. Hangzhou had carried out intensive preparation for the G20 summit and the 2016 event draw the attention of domestic and international media. Hangzhou introduced intense endeavours to improve the environmental aspects in the city (i.e. decrease pollution), and revitalise part of the inner city areas, and develop riverside areas along the Qiantang River.

Moreover, the authorities strengthened control of logistics and transportation in the city to smooth the realisation of the G20 event. Hangzhou authorities also implemented social education and volunteer programs directed towards the local community which introduced identity and responsibility policies towards China, the Hangzhou city, and region. Moreover, city lighting projects and regulations on production impacted positively the city perception by the visitors during the summit. All of the aforementioned actions may be treated as part of the city product development endeavours being a part of event marketing strategies identified by Cudny (2020a).

According to the research results presented in Shu-Wei Tsai's chapter,

> G20 summit Has enhanced local pride and identity as citizens were proud of the city's performance and overall improvement. This event helped the city reach its peak concerning its competitiveness to other cities in China because of the improvements to its environment, the decreased pollution, and economic development caused by the event organization.

The image-related legacies of the G20 summit were strengthened by media relations and thus may be included in promotional elements of event marketing strategies distinguished by Cudny (2020a).

It must be stressed that the aforementioned event-related revitalisation projects and positive narrations associated with the G20 summit were accompanied by social control means introduced before and during the event. Some of the city inhabitants were forced to temporarily move outside the city to lower pollution and avoid crowding. The travel was strictly under control during the G20 summit. Moreover, the disputes over revitalisation projects (questioned by some citizens) and the high-costs of event organisation presented in cyberspace were muted by the authorities. These issues should be treated as disadvantages of the G20 organisation. Aforementioned issues may be compared to pathologies revealed in Broudehoux's (2020) publication presenting dark aspects of mega-events organisation in Rio de Janeiro (Brasil) i.e. displacement of people, demolition of inner-city districts, and counter narrations against official event-related propaganda.

The next chapter was written by Yifan Xu and was devoted to the role of the China Shanghai International Arts Festival (CSIAF 2018 edition) in branding of the city of Shanghai, China. Shanghai is a city in East China, at the intersection of the Yangtze River, Huangpu River, and Suzhou River. Shanghai is one of China's major cities inhabited by over 26 million residents as of 2019. Shanghai is also one of the world's largest seaports. It is a city with a long history but at the same time a modern city, known in the world as a large industrial, service, scientific and financial hub. In the city system operating in China, Shanghai is considered as one of the four largest and most important cities in the country.

Shanghai's economy developed intensively in the last three decades. The growth included the establishment of the business and cultural Pudong District, as well as the special economic zone, rise of infrastructure, development of the city's economy. Shanghai was also a host place for international mega-events like the 2010 Shanghai EXPO.

One of the new aims accepted in the current strategy of Shanghai development is the strengthening of the cultural and event sector in the city. The strategy is supposed to be a way of urban development and branding and the adaptation of well-known western urban strategies of culture and event-led development based on creativity and cultural industries.

The city put a lot of effort into the development of cultural infrastructure, national and international cultural exchange including the performing arts, dance, painting, music, and theatre. Many facilities capable of hosting international mega-events were constructed in Shanghai in recent years.

China Shanghai International Arts Festival (CSIAF) is part of these culture-led development endeavours. The festival is hosted by the Ministry of Culture (currently the Ministry of Culture and Tourism) with the cooperation of Shanghai city. CSIAF is an annual festival staged in October and November, including many events presenting performing arts. It includes music, dance, theatre, Chinese opera, and visual arts exhibitions. The participants are regional and Chinese artists and performers, as well as international artists, invited to participate in the festival.

On one side the event is a large cultural festival and may be treated as a place product offering cultural attractions to inhabitants and domestic as well as international tourists. So the event serves as the city's cultural attraction. On the other hand, the CSIAF is involved in the creation of the brand of Shanghai as a global cultural city, through messages presented on government websites, mainstream media, news reports, and commentaries of festival stakeholders, and the festival's official website. Therefore it serves both nation branding and domestic branding agendas. As noted by Yifan Xu in her chapter

> The study suggested that the city branding agenda has been prioritized over national agendas in the 2018 CSIAF, especially in branding towards international audiences. This is due to the increasing autonomy of the city in planning and managing cultural affairs since the 2010 Shanghai EXPO and the rising urban global city agenda.

CSIAF is a famous cultural event attracting local artists as well as guests from other regions of China and from abroad. Moreover, the CSIAF programme encompasses local culture and local as well as regional artists presenting Shanghai, the region, and its cultural heritage and creative potential. Therefore through the lens of the festival, Chinese citizens perceive Shanghai as famous for cultural development. The media representations of CSIAF as well as social media reports and word-of-mouth messages spread and strengthened this city image to the residents of China among an international audience. Eventually, in relation to Cudny (2020a), the CSIAF may be treated as a festival that realises both elements of event marketing strategy i.e. creating event-related place product based on culture and introduces city promotion and narratives branding Shanghai as a global city of culture utilising local and regional resources for its development.

Chapter 4 differs from the previous two parts of the book. In this chapter Nicholas Wise summarised the role of events on the place product development and promotion of countries associated in the Association of Southeast Asian Nations (ASEAN). The chapter is based on literature analysis, author's own observation, and internet search. It offers a comprehensive review of event-related policies introduced across the ASEAN member countries and cities. The author remarked that the event organisation is a good opportunity to distinguish a country against its competitors. Events offer a perfect opportunity based on the unique event experiences to make a host place distinctive. As a result of event organisation a place receives extraordinary place product and if this product is exciting enough and draws media attention it gains event-related promotion opportunities. Based on the in-depth regional analysis the author suggested organising more medium- and small-scale events instead of focusing so strong on mega-events in the region. Small or regional events are also very attractive to the audience. Moreover, these types of events are regularly organised (annually) and thus are not one-time occurrences like most mega-events. Moreover one of the conclusions of Nicholas Wise's chapter was also the advice to focus on MICE events in the ASEAN countries. Asia and the Pacific is the fastest-growing region in the world in terms of economy. Currently, we witness the move of the world's economic centre to this region. Therefore the MICE sector which includes business-related events (like business meetings, conferences, trade fairs, etc.) has great potential in this part of the world. Therefore business-related events should be strategic for ASEAN countries.

Moreover, the chapter concludes that within the presented region certain cities stand out from the others in terms of event organisation. These are for example Singapore, Kuala Lumpur, Jakarta, Bangkok, and Manilla. These are big, capital cities, with large financial resources and well-developed tourist facilities including venues for event organisation. However smaller destinations are also focusing on event organisation as an opportunity for place product development and promotion. ASEAN countries intend to present their unique culture, build international influence, and create business development opportunities by utilising event-led development strategies. There are a plethora of events organised in cities across different countries in ASEAN. They are very interesting, representing different types according to the scale and theme, and they attract crowds of event-goers, including tourists.

However, according to the author of this chapter, there is a lack of regional (ASEAN oriented) event development strategy which would help develop the event sector, avoid unnecessary competition, and promote all member countries and cities across them. The author mentions also the negative results of the COVID-19 crisis which will be distractive to event-led development strategies in the ASEAN countries at least in a short term. On the other side, long-term event development plans may remain a good opportunity for urban and regional development and promotion in the post-COVID world.

Chapter 5 presented the case study of Okinawa Mathon and its influence on the brand of Okinawa prefecture in Japan. The chapter was written by Yosuke Tsuji and Carolin Schlueter. This chapter together with the following one about Le Tour de France Saitama Criterium (by Daichi Oshimi) presented two Japanese case studies of sports events and their influence on the product and brand of places. Both chapters are based on quantitative methods i.e. statistically analysed results of questionnaires distributed among the visitors participating in the events.

Okinawa is a prefecture in southern Japan, encompassing over 100 islands. The prefecture is a very interesting tourist region in Japan and is, therefore, one of the most visited prefectures in the country. The Okinawa Marathon is a sporting event held in February. The event was first organised in 1955 in a different location and was moved to central Okinawa in 1993. The full marathon and the 10-km long run are nowadays the main races of the event. As stated by the authors of this chapter in recent years the marathon attracted between 14,000 and 15,000 participating runners. The event attracts participants from Okinawa (ca. 73 % of runners in 2017), other regions of Japan, and from overseas (ca. 27% of runners in 2017).

The main aims of the chapter about Okinawa Marathon included presenting the brand association dimensions in the participant sport setting, explaining the relationships between attributes, benefits, and attitudes, and the effects of event-related brand association on the brand equity of the host region. Several methods were used to obtain the necessary data and information. The methods included participant observation and an online questionnaire for the marathon participants.

In the case of Okinawa Marathon, the most important elements evoking event brand associations encompassed the uniqueness of the event, comprising both product-related and non-product-related attributes. The presence of other runners (invited runners, runners in costumes, etc.) was also considered as an important asset of the event increasing its branding effects. The challenge and drama created during the run (i.e. due to time limits and competition between participants) also evoked strong emotions and brand associations expressed in the research results. Attributes related to the course of the marathon, additional events associated with the main run, and details of event organisation were important in the creation of the event-related brand association. The study identified psychological, social, and health benefits presented as advantages from event participation by the respondents. The chapter also explained how the event's brand association influenced the event's brand equity and the brand equity of the host place i.e. Okinawa prefecture.

The results confirmed that a good brand image of a sporting event make a positive influence on a host place and evokes revisit intentions to the destination. The Okinawa Marathon in light of the research results presented in this chapter is an important sports tourism product and an event with positive and strong brand associations. The event, thanks to emotions that it evokes, promotes and positively influences the place brand by shaping a positive perception of Okinawa (including its urban locations).

Chapter 6, written by Daichi Oshimi, presented the case study of Le Tour de France Saitama Criterium. It is the second chapter with a case study from Japan, and another chapter (following the previous one about the Okinawa Marathon) based on a quantitative methodology. Saitama is a large Japanese city located near Tokyo and inhabited by 1.25 million people and a centre of manufacturing and services. Saitama is also well equipped with sports facilities like Saitama Stadium, and multipurpose arena Saitama Super Arena. The city was a host of the FIFA World Cup which was organised together by Japan and the Republic of Korea in 2002. The city put strong efforts into strengthening its brand and perception which is currently less favourable than the image of other Japanese big cities such as Tokyo or Yokohama. For example in 2012 the Saitama Sport Commission was established to realise the new strategy for city promotion through sports.

Between 2011 and 2015, Saitama hosted 157 different types of sporting events and was visited by 811,488 event-goers. The economic impact of these events on Saitama city is estimated at 28.7 billion USD. One of the sporting events organised in the city is a cycling race called Le Tour de France Saitama Criterium (further referred to as the Saitama Criterium). The brand Tour de France is franchised by the French organiser of the race. Using the popular brand is a perfect tool for the promotion of the city of Saitama. The event is one day long and commemorates the 100th anniversary of the Tour de France. It is organised by Saitama Sport Commission, Saitama Prefecture, Saitama City, Saitama Tourism and International Relations Bureau, and Amaury Sport Organisation and is planned for the years 2013–2021. The race is accompanied by entertainment programmes presenting French culture, food, wine, and different specialty products.

The chapter author conducted a questionnaire survey among the event spectators during the 2015 Saitama Criterium. The survey was returned by 860 respondents, 70% of whom were from outside the city. The questionnaire regarded the economic, social, and image impacts of the race on the city of Saitama. The research results presented in the chapter proved that the economic impact of the race, estimated at 2.51 billion USD, was important for the city and the region. Moreover as stated by Daichi Oshimi in his chapter the event's

> Social impacts contribute to the positive attitude towards the host city and the event itself. Especially, perceived "economic development", "cultural interest and new opportunity", "city image", and "event excitement" could be influential tools for city promotion as well as for positive behavioural intention towards the host city and event.

The results presented in this chapter again proved that a sports event (i.e. cycling race) could be a successful product attracting spectators (including tourists) to a hot city and offering them an interesting opportunity to spend free time. The famous brand franchised to the race in Saitama was also important in gaining attention. Moreover, such an event has an important economic impact on the city, and the benefits influence positively the image and brand of the host place and create a possibility for its promotion. The results of both chapters presenting case studies from Japan support the place event marketing model elaborated by Cudny (2020a) and characterising place event marketing as a twofold procedure of place product creation and place promotion with the use of events.

Chapter 7 discusses Durga Puja a religious festival organised in Kolkata, India. In this part of the book, Aparajita De analysed how a religious festival can evolve into a creative cultural event and how such an event can contribute to the development of a city brand. Kolkata city is located in the eastern part of India, on the Hooghly River – a distributary of the Ganges River, near the border of Bangladesh. It is the capital of the West Bengal state and the third-largest metropolis in the country with over 14 million inhabitants in the metropolitan area.

Durga Puja is an annual religious festival held in September and October. It is devoted to Durga, the Hindu deity of war, strength, and protection. Durga Puja routs can be traced back to the eighteenth century. Today the festival includes feasting with various cultural performances like dance, singing, mimicry or swang, puppetry, and folk theatre. Moreover, for the festival, Durga Puja Pandals are created. These are temporary structures with presentations on the deity of Durga. Pandals are usually decorated with clay models of the Goddess Durga, accompanied by the Goddess's four children Laxmi and Sarawati, Ganesha and Kartick. The festival is marked by two narratives – one about the battle between the Maa Durga and Mahisasur demon, reflecting the universal fight between god and evil. The second narrative regards the harvesting season in Bengal.

The chapter used a multi-method approach relying on qualitative research methods applied between 2017 and 2019. The methods included literature analysis, the analysis of media coverage regarding the event. Moreover, direct observations were conducted, in-depth interviews, and digital ethnography of Facebook, blogs, YouTube, and Instagram.

The results of the research presented in the chapter indicate that Durga Puja has evolved from a religious festival to a cultural event. Currently, it is a tourist attraction in Kolkata, and an opportunity to present the local culture rooted in religious beliefs. Apart from that, the festival is also a place for presenting artists cultivating local folklore (dance, theater, music). Durga Puja Pandals are very important, as they are a material element that changes the city space for the duration of the event. Today Durga Puja is a place of cultivation of local tradition and culture. It is a festival where artists representing the creative sectors present themselves. The event attracts tourists from the region, from India, and from abroad. Moreover, as Aparajita De points out, the festival is an element of the festivalisation of the urban space of Calcutta (for city festivalisation, see: Cudny 2016). Because it generates material changes in a part of the city space for the duration of the festival. Physical constructions (pandals) are created, the way of using urban space changes, and new interactions between residents and people participating in the event are created.

In this way, Durga Puja is not only a tourist attraction of Kolkata but also creates its space. Besides, Durga Puja is also an important element in creating a narrative about the city. The festival is shown in the media, especially on social media. The analysis presented in the chapter showed that the event arouses great interest among Internet users, generates a large number of posts, videos (e.g. on Facebook and YouTube), and is widely commented on. All of this contributes to strengthening the image and brand of Kolkata as a city of culture and creative industries. Durga Puja is an important event which on the one hand is a tourist value and an urban product related to religion, culture, and creativity, and on the other hand, it is an important element of promoting the city in the media, thus creating the Kolkata brand.

Chapter 8 written by Sanjukta Sattar presents the city of Mumbai (formerly Bombay) in India and the role of its cultural sector, focusing on the film, fashion industry, and event sector, in brand creation and promotion of the city. Mumbai is the capital city of the state of Maharashtra lying in south-west India and an important trade and business centre and a seaport on the Arabian Sea. The city is inhabited by 12.4 million people and is one of the largest cities in India. Mumbai is especially famous for its growing cultural sector and is currently best known as the capital of the Indian film industry referred to as Bollywood. The Bombay film industry is now considered as a multi-million media sector with global reach. The Bollywood movies created its distinctiveness and are popular not only in India but also abroad.

Mumbai is also a capital of the fashion industry and hosts many events like multi-cultural festivals, music concerts, sports events, literature events. The image of the city is therefore currently shaped by its cultural industries. Mumbai is promoted as the Indian film, fashion, and entertainment capital, cultural and creative city. The city is promoted also with the use of various events like film festivals, award ceremonies, fashion shows, and trade shows organised throughout the year. All these events play important role in the promotion of the city and its brand creation.

Sanjukta Sattar's chapter is based on various qualitative methods. The author investigated the opinion of event organisers, participants, and various professionals regarding the branding and promotional role of the film and event sector, and the fashion industry on the perception and brand of Mumbai. Intensive web-search helped to get information about Mumbai's event sector structure. Moreover reports, research publications were analysed in terms of information and data presenting types of events organised in the city in recent years.

The results of the analysis show the strong association between the cultural and entertainment sector (including events) of Mumbai and the city perception and brand creation. Mumbai is perceived as the entertainment capital of India among others thanks to the presence of Bollywood, the famous Hindi film industry. Bollywood gives the city a distinctive image and global recognition. Moreover, the city is regarded as the centre of the fashion industry. Additionally, other cultural sectors like the music sector and the multicultural heritage of the city are also perceived as Mumbai's strengths. Various events are held throughout the year and they are related among others to the influential cultural sectors (i.e. film industry, fashion, music) and are also related to Mumbai's multiculturalism. They enrich the entertainment sector, attract tourists, and help to create the image of Mumbai as India's entertainment capital. Moreover, they promote the film and fashion sectors and help to build the brand of the city as the global centre of the film industry and fashion capital. Again the chapters with case-studies from India supported the thesis of Cudny (2020a) regarding the place event marketing and the structure of the process encompassing place product creation and urban promotion.

The last factual chapter (Chapter 9) of this edited volume is devoted to America's Cup ocean race and its impact on the brand of the city of Auckland, New Zealand. Richard Keith Wright and Christopher Barron focused their chapter on the perceived legacies of the large ocean race that occured in Auckland in March 2021. The chapter is different from the previously presented parts of the book because it estimates the branding impacts of a future event. However, it is also related to the other chapters, e.g. to case studies from Japan because it presents an analysis of a sporting event. The text is based on qualitative methodology, like most of the contributions in this book, methods include in-depth literature and media analysis.

Auckland is New Zealand's largest city, located on the North Island between the Tasman Sea and the Pacific Ocean. Auckland is an important economic, cultural, and scientific center and a major seaport in the country. America's Cup regatta (also known as Auld Mug) is an international sailing race held since 1851. It is currently one of the world's most known and valued ocean races. The history of America's Cup began in 1851 with the victory of the American yacht during the regatta around the Isle of Wight. In 2021, the Royal New Zealand Yacht Squadron will become the first non-American sailing club to host an America's Cup regatta on more than two occasions (previously in 2000 and 2003).

As stated by the chapter's authors the upcoming ocean race is a powerful tool in the creation of Auckland's international perceptions, and the city brand identity. It is a great opportunity to reinforce Auckland's brand positioning as the "City of Sails" and as a world-leading sport tourist destination. Moreover, the city authorities see race as the opportunity to continue the downtown waterfront's revitalisation, including the creation of new land and water spaces serving for America's Cup 2021 and future water-based events in Auckland. These investments together with the hosting of Auld Mug 2021 will enhance Auckland's chances to bid for large international water sports events in the future. The authors of the chapter argue that America's Cup 2021 will draw a global audience during and after the event. It is predicted that the ocean race will bring important benefits in terms of economy and employment to the city and region of Auckland.

The event will also help to promote Auckland, a remote tourist destination, in international media and on social media as an important host city for major water sporting events. This is in line with the strategic plans of the city regarding the development of tourism, including sports tourism based on water sports and sailing. It must also be stressed that the disruption induced by the spread of the COVID-19 virus forms a serious threat to the realisation of America's Cup in 2021. However, as noticed by the authors the latest discovery of vaccines and their planned delivery in early 2021 may prevent the further spread of the pandemic, and Spring 2021 could be a start in the post-pandemic world. In this case, America's Cup could be realised and maybe a great success, even though it still will have to be under a strict sanitary regime. Eventually, if realised, America's Cup 2021 in Auckland may be a successful kickstart in New Zealand's and Auckland's tourism sector recovery and a successful tool for city branding and promotion with the use of an event.

This book is the second publication devoted to the issue of place evet marketing edited by Waldemar Cudny. This edited volume presents a series of interesting case studies showing the role of events in urban product development and in promoting cities in Asia and the Pacific region. It should be emphasised, however, that the research presented in the book has its limitations. To fully confirm the role of events in creating the development of urban products and promoting cities, further research in the field of place event marketing is necessary.

The analysis presented in this book is based on selected examples. Therefore, the research on place event marketing should be supplemented with further case studies from different parts of the world. Besides, it would be valuable to test the model describing place event marketing presented by Cudny (2020a) and also verified in this book on further examples. Subsequent research covering various (in terms of size, subject, and location) events would allow verifying the model's correctness and scientific value.

Moreover, it should be emphasised that the year 2020 brought the global health and socio-economic crisis caused by the COVID-19 pandemic. This crisis also affected the tourism industry and the event sector cooperating with it. Undoubtedly, the effects of the pandemic on tourism and the organisation of events are very negative. Limitation of the organisation of events as well as strict sanitary procedures hindered the benefits resulting from the organisation of events including their social, economic, and promotional legacies. On the other hand, widespread testing and the discovery of a COVID-19 vaccine will give an end to the pandemic in due time. Therefore, cities will be again centers of tourism and places of development of the experience-based industries such as the event sector. They will look for a new start in a post-pandemic reality, and revive products that regenerate the economy, entertain and attract tourists. In the post-covid era, events will certainly once again become elements of place event marketing strategies creating urban experience-based attractions and enabling cities to promote themselves.

The analysis presented in the edited volume revealed that the model proposed in the publication Cudny (2020a, 17) is confirmed by empirical studies conducted in the area of selected cities in the Asia Pacific region. The research presented and summarised above, being case studies from China, India, Japan, New Zealand, and ASEAN countries, showed a significant role of events in creating urban products and in promoting host cities and towns. The model was confirmed by the example of events of various sizes (from mega-events to regional events) and themes (cultural, religious, sports, and MICE events). In addition, it was shown that the purpose of organising events that were supported by city authorities, the central government, and many non-governmental organisations was to purposefully (strategically) create the development of cities using sports, cultural and entertainment events, and business-related MICE events. The events were also part of long-term plans for socio-economic development and promotion plans introduced in some of the cities described in the book. In the case of most of the events described in this edition, there were many positive impacts on cities. These impacts can be described in reference to the literature as positive event legacies or impacts (see: Preuss 2007; Smith 2012; Cudny 2016 (Table 10.1).

These legacies included the creation of events as experience-based urban products and development of urban space (often referred to as festivalisation of urban spaces) for instance by tangible investments (in infrastructure, venues, etc.), revitalisation programmes, creation of contemporary event-related constructions (e.g. for Durga Puja festival in Kolkata). Other legacies of events presented in this book encompassed the growth of tourism (MICE, festival, cultural and religious, and sports tourism) and the creation of cultural and entertainment offers for local visitors. Some events were encompassed with social programmes including volunteer training or education. The aforementioned impacts of events may be treated as part of policies of urban product development through events, depicted as part of place event marketing in Cudny's (2020a, 17) model (see: Table 10.1).

Events also evoked diverse promotional legacies. They could be treated as part of the second type of activities forming place event marketing i.e. these connected with place promotion and city brand creation (Cudny 2020a, 17). The events from the Asia Pacific region presented in this book evoked such promotional effects as promotion of the host destinations in the media (traditional as well as social media), promotion of the host country in relation to the event. Moreover,

Table 10.1 Legacies of events from the Asia Pacific region presented through the book

Results of place event marketing	Cities and towns	Name of the event/events
Creation of experience-based urban products	Hangzhou (China)	G20 summit
	Shanghai (China)	China Shanghai International Arts Festival (CSIAF)
	ASEAN countries	Different events
	Okinawa (Japan)	Okinawa Mathon
	Saitama (Japan)	Le Tour de France Saitama Criterium
	Kolkata (India)	Durga Puja
	Mumbai (India)	Event and creative industries
	Auckland (New Zealand)	America's Cup
Urban development through event-related tangible investments, revitalisation, and temporary constructions (festivalisation of urban spaces see: Cudny 2016)	Hangzhou (China)	G20 summit
	Kolkata (India)	Durga Puja
	Auckland (New Zealand)	America's Cup
Growth of tourism	Hangzhou (China) – MICE tourism	G20 summit
	Shanghai (China) – cultural tourism	CSIAF
	ASEAN countries – different types of tourism	Different events
	Okinawa (Japan) – sports tourism	Okinawa Marathon
	Saitama (Japan) – sports tourism	Le Tour de France Saitama Criterium
	Kolkata (India) – cultural and religious tourism	Durga Puja
	Mumbai (India) – cultural tourism	Event and creative industries
	Auckland (New Zealand) - sports tourism growth projected for 2021	America's Cup
Creation of a cultural and entertainment offer for local visitors	Hangzhou (China) – via media relations	G20 summit
	Shanghai (China)	CSIAF
	Kolkata (India)	Durga Puja
	Mumbai (India)	Event and creative industries
Introduction of social programmes (e.g. volunteer or education) engaging local inhabitants	Hangzhou (China)	G20 summit

(Continued)

Table 10.1 (Continued)

Results of place event marketing	Cities and towns	Name of the event/ events
Promotion of the host destination in traditional media (TV, press, radio, etc.)	Hangzhou (China)	G20 summit
	Shanghai (China)	CSIAF
	ASEAN countries	Different events
	Okinawa (Japan)	Okinawa Marathon
	Saitama (Japan)	Le Tour de France Saitama Criterium
	Kolkata (India)	Durga Puja
	Mumbai (India)	Event and creative industries
	Auckland (New Zealand)	America's Cup
Promotion of the host destination in social media (e.g. on Facebook, Youtube, blogs, internet websites)	Hangzhou (China) – associated with strong governmental control over the media content	G20 summit
	Shanghai (China)	CSIAF
	ASEAN countries	Different events
	Okinawa (Japan)	Okinawa Marathon
	Saitama (Japan)	Le Tour de France Saitama Criterium
	Kolkata (India)	Durga Puja
	Mumbai (India)	Event and creative industries
	Auckland (New Zealand)	America's Cup
Promotion of the host country concerning the event organised in a city	Hangzhou (China)	G20 summit
	Shanghai (China)	CSIAF
	Auckland (New Zealand)	America's Cup
Strengthening of city /town image in the eyes of external recipients (e.g. domestic and international tourists)	Hangzhou (China)	G20 summit
	Shanghai (China)	CSIAF
	ASEAN countries	Different events
	Okinawa (Japan)	Okinawa Marathon
	Saitama (Japan)	Le Tour de France Saitama Criterium
	Kolkata (India)	Durga Puja
	Mumbai (India)	Event and creative industries
	Auckland (New Zealand)	America's Cup
Strengthening of city/town image in the eyes of internal recipients (urban dwellers)	Hangzhou (China)	G20 summit
	Shanghai (China)	CSIAF
	Okinawa (Japan)	Okinawa Marathon
	Saitama (Japan)	Le Tour de France Saitama Criterium
	Kolkata (India)	Durga Puja
	Mumbai (India)	Event and creative industries
	Auckland (New Zealand)	America's Cup

Source: Author's elaboration.

strengthening of the host place image was noticed both by tourists and city dwellers (Table 10.1).

References

Broudehoux, A.M. (2017). *Mega Events and Urban Image Construction: Bejing and Rio de Janeiro*. London and New York: Routledge.
Broudehoux, A.M. (2020). Resisting Rio de Janeiro's event-led place promotion: From insurgent rebranding to festive counter-spectacle. In: W. Cudny ed. *Urban Events, Place Branding and Promotion. Place Event Marketing*, pp. 124–140. London: Routledge.
Cudny, W. (2016). *Festivalisation of Urban Spaces: Factors, Processes and Effects*. Cham: Springer.
Cudny, W. (2019). *City Branding and Promotion: The Strategic Approach*. London and New York: Routledge.
Cudny, W. (2020a) The concept of place event marketing: Setting the agenda. In: W. Cudny ed. *Urban Events, Place Branding and Promotion. Place Event Marketing*, pp. 1–24. London-New York: Routledge.
Cudny, W. (ed.) (2020b) *Urban Events, Place Branding and Promotion. Place Event Marketing*. London and New York: Routledge.
Degen, M., & Garcia, M. (2012) The transformation of the 'Barcelona model': An analysis of culture, urban regeneration and governance. *International Journal of Urban and Regional Research*, 36 (5), 1022–1038.
Fang, Z., Chang, Y. (2016) Energy, human capital and economic growth in Asia Pacific countries: Evidence from a panel cointegration and causality analysis. *Energy Economics*, 56, 177–184.
Hoyle, L.H. (2002) *Event marketing. How to Successfully Promote Events, Festivals, Conventions and Expositions*. New York: Wiley.
International Tourism Highlights (2019) *UNWTO World Tourism Organization*, Edition 2019, available at: https://www.e-unwto.org/doi/pdf/10.18111/9789284421152, accessed on 18.08.2020.
Jaworowicz, M., & Jaworowicz, P. (2016) *Event marketing w zintegrowanej komunikacji marketingowej*. Warszawa, SA: Difin.
Kotler, P., Armstrong, G. (2010) *Principles of Marketing*. Upper Saddle River: Pearson, Prentice Hall.
Preuss, H. (2007). The conceptualisation and measurement of mega sport event legacies. *Journal of Sport & Tourism*, 12(3–4), 207–228.
Smith, A.(2012) *Events and Urban Regeneration: The Strategic Use of Events to Revitalise Cities*. London and New York: Routledge.
Tolkach, D., Chon, K.K., & Xiao, H. (2016). Asia Pacific tourism trends: Is the future ours to see? *Asia Pacific Journal of Tourism Research*, 21 (10), 1071–1084.
Wilson, R. (2004) The impact of cultural events on city image: Rotterdam, cultural capital of Europe 2001. *Urban Studies*, 41 (10), 1993.

Index

Page numbers in **bold** denote tables, those in *italic* denote figures.

For Product Safety Concerns and Information please contact our EU
representative GPSR@taylorandfrancis.com
Taylor & Francis Verlag GmbH, Kaufingerstraße 24, 80331 München, Germany

www.ingramcontent.com/pod-product-compliance
Lightning Source LLC
Chambersburg PA
CBHW060305220326
41598CB00027B/4239